Wonderful Life

W · W · NORTON & COMPANY · NEW YORK · LONDON

Wonderful Life

The Burgess Shale and the Nature of History

STEPHEN JAY GOULD

First published as a Norton paperback 1990

The text of this book is composed in 10½/13 Avanta, with
display type set in Fenice Light. Composition and
manufacturing by The Haddon Craftsmen, Inc.
Book design by Antonina Krass.

Library of Congress Cataloging-in-Publication Data
Gould, Stephen Jay.
Wonderful life: the Burgess Shale and the nature of history / Stephen Jay Gould.
p. cm. Bibliography: p. Includes index.
1. Evolution—History. 2. Invertebrates, Fossil. 3. Paleontology—Cambrian.
4. Paleontology—British Columbia—Yoho National Park. 5. Burgess
Shale. 6. Paleontology—Philosophy. 7. Contingency (Philosophy)
8. Yoho National Park (B.C.) I. Title.
QE770.G67 1989
560'.9—dc19   88-37469

ISBN 0-393-30700-X

W. W. Norton & Company, Inc., 500 Fifth Avenue, New York, N. Y. 10110
W. W. Norton & Company Ltd., 10 Coptic Street, London WC1A 1PU
8 9 0

To Norman D. Newell

Who was, and is, in the most noble
word of all human speech, my teacher

Contents

Preface and Acknowledgments

This book, to cite some metaphors from my least favorite sport, attempts to tackle one of the broadest issues that science can address—the nature of history itself—not by a direct assault upon the center, but by an end run through the details of a truly wondrous case study. In so doing, I follow the strategy of all my general writing. Detail by itself can go no further; at its best, presented with a poetry that I cannot muster, it emerges as admirable "nature writing." But frontal attacks upon generalities inevitably lapse into tedium or tendentiousness. The beauty of nature lies in detail; the message, in generality. Optimal appreciation demands both, and I know no better tactic than the illustration of exciting principles by well-chosen particulars.

My specific topic is the most precious and important of all fossil localities—the Burgess Shale of British Columbia. The human story of discovery and interpretation, spanning almost eighty years, is wonderful, in the strong literal sense of that much-abused word. Charles Doolittle Walcott, premier paleontologist and most powerful administrator in American science, found this oldest fauna of exquisitely preserved soft-bodied animals in 1909. But his deeply traditionalist stance virtually forced a conventional interpretation that offered no new perspective on life's history, and therefore rendered these unique organisms invisible to public notice (though they far surpass dinosaurs in their potential for instruction about life's history). But twenty years of meticulous anatomical description by three

English and Irish paleontologists, who began their work with no inkling of its radical potential, has not only reversed Walcott's interpretation of these particular fossils, but has also confronted our traditional view about progress and predictability in the history of life with the historian's challenge of contingency—the "pageant" of evolution as a staggeringly improbable series of events, sensible enough in retrospect and subject to rigorous explanation, but utterly unpredictable and quite unrepeatable. Wind back the tape of life to the early days of the Burgess Shale; let it play again from an identical starting point, and the chance becomes vanishingly small that anything like human intelligence would grace the replay.

But even more wonderful than any human effort or revised interpretation are the organisms of the Burgess Shale themselves, particularly as newly and properly reconstructed in their transcendent strangeness: *Opabinia*, with its five eyes and frontal "nozzle"; *Anomalocaris*, the largest animal of its time, a fearsome predator with a circular jaw; *Hallucigenia*, with an anatomy to match its name.

The title of this book expresses the duality of our wonder—at the beauty of the organisms themselves, and at the new view of life that they have inspired. *Opabinia* and company constituted the strange and wonderful life of a remote past; they have also imposed the great theme of contingency in history upon a science uncomfortable with such concepts. This theme is central to the most memorable scene in America's most beloved film—Jimmy Stewart's guardian angel replaying life's tape without him, and demonstrating the awesome power of apparent insignificance in history. Science has dealt poorly with the concept of contingency, but film and literature have always found it fascinating. *It's a Wonderful Life* is both a symbol and the finest illustration I know for the cardinal theme of this book—and I honor Clarence Odbody, George Bailey, and Frank Capra in my title.

The story of the reinterpretation of the Burgess fossils, and of the new ideas that emerged from this work, is complex, involving the collective efforts of a large cast. But three paleontologists dominate the center stage, for they have done the great bulk of technical work in anatomical description and taxonomic placement—Harry Whittington of Cambridge University, the world's expert on trilobites, and two men who began as his graduate students and then built brilliant careers upon their studies of the Burgess fossils, Derek Briggs and Simon Conway Morris.

I struggled for many months over various formats for presenting this work, but finally decided that only one could provide unity and establish integrity. If the influence of history is so strong in setting the order of life today, then I must respect its power in the smaller domain of this book.

The work of Whittington and colleagues also forms a history, and the primary criterion of order in the domain of contingency is, and must be, chronology. The reinterpretation of the Burgess Shale is a story, a grand and wonderful story of the highest intellectual merit—with no one killed, no one even injured or scratched, but a new world revealed. What else can I do but tell this story in proper temporal order? Like *Rashomon,* no two observers or participants will ever recount such a complex tale in the same manner, but we can at least establish a groundwork in chronology. I have come to view this temporal sequence as an intense drama—and have even permitted myself the conceit of presenting it as a play in five acts, embedded within my third chapter.

Chapter I lays out, through the unconventional device of iconography, the traditional attitudes (or thinly veiled cultural hopes) that the Burgess Shale now challenges. Chapter II presents the requisite background material on the early history of life, the nature of the fossil record, and the particular setting of the Burgess Shale itself. Chapter III then documents, as a drama and in chronological order, this great revision in our concepts about early life. A final section tries to place this history in the general context of an evolutionary theory partly challenged and revised by the story itself. Chapter IV probes the times and psyche of Charles Doolittle Walcott, in an attempt to understand why he mistook so thoroughly the nature and meaning of his greatest discovery. It then presents a different and antithetical view of history as contingency. Chapter V develops this view of history, both by general arguments and by a chronology of key episodes that, with tiny alterations at the outset, could have sent evolution cascading down wildly different but equally intelligible channels—sensible pathways that would have yielded no species capable of producing a chronicle or deciphering the pageant of its past. The epilogue is a final Burgess surprise—*vox clamantis in deserto,* but a happy voice that will not make the crooked straight or the rough places plain, because it revels in the tortuous crookedness of real paths destined only for interesting ends.

I am caught between the two poles of conventional composition. I am not a reporter or "science writer" interviewing people from another domain under the conceit of passive impartiality. I am a professional paleontologist, a close colleague and personal friend of all the major actors in this drama. But I did not perform any of the primary research myself—nor could I, for I do not have the special kind of spatial genius that this work requires. Still, the world of Whittington, Briggs, and Conway Morris is my world. I know its hopes and foibles, its jargon and techniques, but I also live with its illusions. If this book works, then I have combined a professional's feeling and knowledge with the distance necessary for judgment, and my dream of writing an "insider's McPhee" within geology may have succeeded. If it does not work, then I am simply the latest of so many victims—and all the clichés about fish and fowl, rocks and hard places, apply. (My difficulty in simultaneously living in and reporting about this world emerges most frequently in a simple problem that I found insoluble. Are my heroes called Whittington, Briggs, and Conway Morris; or are they Harry, Derek, and Simon? I finally gave up on consistency and decided that both designations are appropriate, but in different circumstances— and I simply followed my instinct and feeling. I had to adopt one other convention; in rendering the Burgess drama chronologically, I followed the dates of publication for ordering the research on various Burgess fossils. But as all professionals know, the time between manuscript and print varies capriciously and at random, and the sequence of publication may bear little relationship to the order of actual work. I therefore vetted my sequence with all the major participants, and learned, with pleasure and relief, that the chronology of publication acted as a pretty fair surrogate for order of work in this case.)

I have fiercely maintained one personal rule in all my so-called "popular" writing. (The word is admirable in its literal sense, but has been debased to mean simplified or adulterated for easy listening without effort in return.) I believe—as Galileo did when he wrote his two greatest works as dialogues in Italian rather than didactic treatises in Latin, as Thomas Henry Huxley did when he composed his masterful prose free from jargon, as Darwin did when he published all his books for general audiences—that we can still have a genre of scientific books suitable for and accessible alike to professionals and interested laypeople. The concepts of science, in all their richness and ambiguity, can be presented without any compromise, without any simplification counting as distortion, in language accessible to all intelligent people. Words, of course, must be varied, if only to eliminate a jargon and phraseology that would mystify anyone outside the priesthood, but conceptual depth should not vary at all between professional publication and general exposition. I hope that this book can be read with profit both in seminars for graduate students and—if the movie stinks and you forgot your sleeping pills—on the businessman's special to Tokyo.

Of course, these high-minded hopes and conceits from yours truly also demand some work in return. The beauty of the Burgess story lies in its details, and the details are anatomical. Oh, you could skip the anatomy and still get the general message (Lord knows, I repeat it enough times in my enthusiasm)—but please don't, for you will then never understand either the fierce beauty or the intense excitement of the Burgess drama. I have done everything I could to make the two technical subjects—anatomy and taxonomy—maximally coherent and minimally intrusive. I have provided insets as primers on these subjects, and I have kept the terminology to an absolute minimum (fortunately, we can bypass nearly all the crushing jargon of professional lingo, and grasp the key point about arthropods by simply understanding a few facts about the order and arrangement of appendages). In addition, all descriptive statements in the text are matched by illustrations.

I did briefly consider (but it was only the Devil speaking) the excision of all this documentation, with a bypass via some hand waving, pretty pictures, and an appeal to authority. But I could not do it—and not only for reasons of general policy mentioned above. I could not do it because any expunging of anatomical arguments, any derivative working from secondary sources rather than primary monographs, would be a mark of disrespect for something truly beautiful—for some of the most elegant technical work ever accomplished in my profession, and for the exquisite loveliness of the Burgess animals. Pleading is undignified, but allow me one line: please bear with the details; they are accessible, and they are the gateway to a new world.

A work like this becomes, perforce, something of a collective enterprise—and thanks for patience, generosity, insight, and good cheer must be widely spread. Harry Whittington, Simon Conway Morris, and Derek Briggs endured hours of interviews, detailed questioning, and reading of manuscripts. Steven Suddes, of Yoho National Park, kindly organized a hike to the hallowed ground of Walcott's quarry, for I could not write this book without making such a pilgrimage. Laszlo Meszoly prepared charts and diagrams with a skill that I have admired and depended upon for nearly two decades. Libby Glenn helped me wade through the voluminous Walcott archives in Washington.

Never before have I published a work so dependent upon illustrations. But so it must be; primates are visual animals above all, and anatomical work, in particular, is as much pictorial as verbal. I decided right at the outset that most of my illustrations must be those originally used in the basic publications of Whittington and colleagues—not only for their excellence within the genre, but primarily because I know no other way to express my immense respect for their work. In this sense, I am only acting as a faithful chronicler of primary sources that will become crucial in the history of my profession. With the usual parochialism of the ignorant, I assumed that the photographic reproduction of published figures must be a simple and automatic procedure of shoot 'em and print 'em. But I learned a lot about other professional excellences as I watched Al Coleman and David Backus, my photographer and my research assistant, work for three months to achieve resolutions that I couldn't see in the primary publications themselves. My greatest thanks for their dedication and their instruction.

These figures—about a hundred, all told—are primarily of two types: drawings of actual specimens, and schematic reconstructions of entire organisms. I could have whited out the labeling of features, often quite dense, on the drawings of specimens, for few of these labels relate to arguments made in my text and those that do are always fully explained in my captions. But I wanted readers to see these illustrations exactly as they appear in the primary sources. Readers should note, by the way, that the reconstructions, following a convention in scientific illustration, rarely show an animal as an observer might have viewed it on a Cambrian sea bottom—and for two reasons. Some parts are usually made transparent, so that more of the full anatomy may be visualized; while other parts (usually those repeated on the other side of the body) are omitted for the same reason.

Since the technical illustrations do not show an organism as a truly living creature, I decided that I must also commission a series of full reconstructions by a scientific artist. I was not satisfied with any of the standard published illustrations—they are either inaccurate or lacking in aesthetic oomph. Luckily, Derek Briggs showed me Marianne Collins's drawing of *Sanctacaris* (figure 3.55), and I finally saw a Burgess organism drawn with a scrupulous attention to anatomical detail combined with aesthetic flair that reminded me of the inscription on the bust of Henry Fairfield Osborn at the American Museum of Natural History: "For him the dry bones came to life, and giant forms of ages past rejoined the pageant of the living." I am delighted that Marianne Collins, of the Royal Ontario Museum, Toronto, was able to provide some twenty drawings of Burgess animals exclusively for this book.

This collective work binds the generations. I spoke extensively with Bill Schevill, who quarried with Percy Raymond in the 1930s, and with G. Evelyn Hutchinson, who published his first notable insights on Burgess fossils just after Walcott's death. Having nearly touched Walcott himself, I ranged to the present and spoke with all active workers. I am especially grateful to Desmond Collins, of the Royal Ontario Museum, who in the summer of 1988, as I wrote this book, was camped in Walcott's original quarry while making fresh discoveries at a new site above Raymond's quarry. His work will expand and revise several sections of my text; obsolescence is a fate devoutly to be wished, lest science stagnate and die.

I have been obsessed with the Burgess Shale for more than a year, and have talked incessantly about its problems with colleagues and students far and wide. Many of their suggestions, and their doubts and cautions, have greatly improved this book. Scientific fraud and general competitive nastiness are hot topics this season. I fear that outsiders are getting a false view of this admittedly serious phenomenon. The reports are so prominent that one might almost envision an act of chicanery for each ordinary event of decency and honor. No, not at all. The tragedy is not the frequency of such acts, but the crushing asymmetry that permits any rare event of unkindness to nullify or overwhelm thousands of collegial gestures, never recorded because we take them for granted. Paleontology is a genial profession. I do not say that we all like each other; we certainly do not agree about very much. But we do tend to be helpful to each other, and to avoid pettiness. This grand tradition has eased the path of this book, through a thousand gestures of kindness that I never recorded because they are the ordinary acts of decent people—that is, thank goodness, most of us most of the time. I rejoice in this sharing, in our joint love for knowledge about the history of our wonderful life.

Wonderful Life

CHAPTER I

The Iconography of an Expectation

A PROLOGUE IN PICTURES

And I will lay sinews upon you, and will bring up flesh upon
you, and cover you with skin, and put breath in you, and ye
shall live.—Ezekiel 37:6

Not since the Lord himself showed his stuff to Ezekiel in the valley
of dry bones had anyone brought such grace and skill to the reconstruction of animals from disarticulated skeletons. Charles R. Knight, most
celebrated of artists in the reanimation of fossils, painted all the canonical
figures of dinosaurs that fire our fear and imagination to this day. In February 1942, Knight designed a chronological series of panoramas, depicting
the history of life from the advent of multicellular animals to the triumph
of *Homo sapiens,* for the *National Geographic.* (This is the one issue that's
always saved and therefore always missing when you see a "complete" run
of the magazine on sale for two bits an issue on the back shelves of the
general store in Bucolia, Maine.) He based his first painting in the series—
shown on the jacket of this book—on the animals of the Burgess Shale.

Without hesitation or ambiguity, and fully mindful of such paleontological wonders as large dinosaurs and African ape-men, I state that the invertebrates of the Burgess Shale, found high in the Canadian Rockies in
Yoho National Park, on the eastern border of British Columbia, are the
world's most important animal fossils. Modern multicellular animals make

their first uncontested appearance in the fossil record some 570 million years ago—and with a bang, not a protracted crescendo. This "Cambrian explosion" marks the advent (at least into direct evidence) of virtually all major groups of modern animals—and all within the minuscule span, geologically speaking, of a few million years. The Burgess Shale represents a period just after this explosion, a time when the full range of its products inhabited our seas. These Canadian fossils are precious because they preserve in exquisite detail, down to the last filament of a trilobite's gill, or the components of a last meal in a worm's gut, the soft anatomy of organisms. Our fossil record is almost exclusively the story of hard parts. But most animals have none, and those that do often reveal very little about their anatomies in their outer coverings (what could you infer about a clam from its shell alone?). Hence, the rare soft-bodied faunas of the fossil record are precious windows into the true range and diversity of ancient life. The Burgess Shale is our only extensive, well-documented window upon that most crucial event in the history of animal life, the first flowering of the Cambrian explosion.

The story of the Burgess Shale is also fascinating in human terms. The fauna was discovered in 1909 by America's greatest paleontologist and scientific administrator, Charles Doolittle Walcott, secretary (their name for boss) of the Smithsonian Institution. Walcott proceeded to misinterpret these fossils in a comprehensive and thoroughly consistent manner arising directly from his conventional view of life: In short, he shoehorned every last Burgess animal into a modern group, viewing the fauna collectively as a set of primitive or ancestral versions of later, improved forms. Walcott's work was not consistently challenged for more than fifty years. In 1971, Professor Harry Whittington of Cambridge University published the first monograph in a comprehensive reexamination that began with Walcott's assumptions and ended with a radical interpretation not only for the Burgess Shale, but (by implication) for the entire history of life, including our own evolution.

This book has three major aims. It is, first and foremost, a chronicle of the intense intellectual drama behind the outward serenity of this reinterpretation. Second, and by unavoidable implication, it is a statement about the nature of history and the awesome improbability of human evolution. As a third theme, I grapple with the enigma of why such a fundamental program of research has been permitted to pass so invisibly before the public gaze. Why is *Opabinia*, key animal in a new view of life, not a household name in all domiciles that care about the riddles of existence?

In short, Harry Whittington and his colleagues have shown that most Burgess organisms do not belong to familiar groups, and that the creatures from this single quarry in British Columbia probably exceed, in anatomical range, the entire spectrum of invertebrate life in today's oceans. Some fifteen to twenty Burgess species cannot be allied with any known group, and should probably be classified as separate phyla. Magnify some of them beyond the few centimeters of their actual size, and you are on the set of a science-fiction film; one particularly arresting creature has been formally named *Hallucigenia*. For species that can be classified within known phyla, Burgess anatomy far exceeds the modern range. The Burgess Shale includes, for example, early representatives of all four major kinds of arthropods, the dominant animals on earth today—the trilobites (now extinct), the crustaceans (including lobsters, crabs, and shrimp), the chelicerates (including spiders and scorpions), and the uniramians (including insects). But the Burgess Shale also contains some twenty to thirty kinds of arthropods that cannot be placed in any modern group. Consider the magnitude of this difference: taxonomists have described almost a million species of arthropods, and all fit into four major groups; one quarry in British Columbia, representing the first explosion of multicellular life, reveals more than twenty additional arthropod designs! The history of life is a story of massive removal followed by differentiation within a few surviving stocks, not the conventional tale of steadily increasing excellence, complexity, and diversity.

For an epitome of this new interpretation, compare Charles R. Knight's restoration of the Burgess fauna (figure 1.1), based entirely on Walcott's classification, with one that accompanied a 1985 article defending the reversed view (figure 1.2).

1. The centerpiece of Knight's reconstruction is an animal named *Sidneyia*, largest of the Burgess arthropods known to Walcott, and an ancestral chelicerate in his view. In the modern version, *Sidneyia* has been banished to the lower right, its place usurped by *Anomalocaris*, a two foot-terror of the Cambrian seas, and one of the Burgess "unclassifiables."

2. Knight restores each animal as a member of a well-known group that enjoyed substantial later success. *Marrella* is reconstructed as a trilobite, *Waptia* as a proto-shrimp (see figure 1.1), though both are ranked among the unplaceable arthropods today. The modern version features the unique phyla—giant *Anomalocaris; Opabinia* with its five eyes and frontal "nozzle"; *Wiwaxia* with its covering of scales and two rows of dorsal spines.

3. Knight's creatures obey the convention of the "peaceable kingdom." All are crowded together in an apparent harmony of mutual toleration; they do not interact. The modern version retains this unrealistic crowding (a necessary tradition for economy's sake), but features the ecological relations uncovered by recent research: priapulid and polychaete worms burrow in the mud; the mysterious *Aysheaia* grazes on sponges; *Anomalocaris*

everts its jaw and crunches a trilobite.

4. Consider *Anomalocaris* as a prototype for Whittington's revision. Knight includes two animals omitted from the modern reconstruction: jellyfish and a curious arthropod that appears to be a shrimp's rear end covered in front by a bivalved shell. Both represent errors committed in the overzealous attempt to shoehorn Burgess animals into modern groups. Walcott's "jellyfish" turns out to be the circlet of plates surrounding the mouth of *Anomalocaris;* the posterior of his "shrimp" is a feeding appendage of the same carnivorous beast. Walcott's prototypes for two modern groups become body parts of the largest Burgess oddball, the appropriately named *Anomalocaris.*

Thus a complex shift in ideas is epitomized by an alteration in pictures. Iconography is a neglected key to changing opinions, for the history and meaning of life in general, and for the Burgess Shale in stark particulars.

1.1.   Reconstruction of the Burgess Shale fauna done by Charles R. Knight in 1940, probably the model for his 1942 restoration. All the animals are drawn as members of modern groups. Above *Sidneyia,* the largest animal of the scene, *Waptia* is reconstructed as a shrimp. Two parts that really belong to the unique creature *Anomalocaris* are portrayed respectively as an ordinary jellyfish (top, left of center) and the rear end of a bivalved arthropod (the large creature, center right, swimming above the two trilobites).

1.2. A modern reconstruction of the Burgess Shale fauna, illustrating an article by Briggs and Whittington on the genus *Anomalocaris*. This drawing, unlike Knight's, features odd organisms. *Sidneyia* has been banished to the lower right, and the scene is dominated by two specimens of the giant *Anomalocaris*. Three *Aysheaia* feed on sponges along the lower border, left of *Sidneyia*. An *Opabinia* crawls along the bottom just left of *Aysheaia*. Two *Wiwaxia* graze on the sea floor below the upper *Anomalocaris*.

THE LADDER AND THE CONE:
ICONOGRAPHIES OF PROGRESS

Familiarity has been breeding overtime in our mottoes, producing everything from contempt (according to Aesop) to children (as Mark Twain observed). Polonius, amidst his loquacious wanderings, urged Laertes to seek friends who were tried and true, and then, having chosen well, to "grapple them" to his "soul with hoops of steel."

Yet, as Polonius's eventual murderer stated in the most famous soliloquy of all time, "there's the rub." Those hoops of steel are not easily unbound, and the comfortably familiar becomes a prison of thought.

Words are our favored means of enforcing consensus; nothing inspires orthodoxy and purposeful unanimity of action so well as a finely crafted motto—Win one for the Gipper, and God shed his grace on thee. But our recent invention of speech cannot entirely bury an earlier heritage. Primates are visual animals par excellence, and the iconography of persuasion strikes even closer than words to the core of our being. Every demagogue, every humorist, every advertising executive, has known and exploited the evocative power of a well-chosen picture.

Scientists lost this insight somewhere along the way. To be sure, we use pictures more than most scholars, art historians excepted. *Next slide please* surpasses even *It seems to me that* as the most common phrase in professional talks at scientific meetings. But we view our pictures only as ancillary illustrations of what we defend by words. Few scientists would view an image itself as intrinsically ideological in content. Pictures, as accurate mirrors of nature, just are.

I can understand such an attitude directed toward photographs of objects—though opportunities for subtle manipulation are legion even here. But many of our pictures are incarnations of concepts masquerading as neutral descriptions of nature. These are the most potent sources of conformity, since ideas passing as descriptions lead us to equate the tentative with the unambiguously factual. Suggestions for the organization of thought are transformed to established patterns in nature. Guesses and hunches become things.

The familiar iconographies of evolution are all directed—sometimes crudely, sometimes subtly—toward reinforcing a comfortable view of human inevitability and superiority. The starkest version, the chain of being or ladder of linear progress, has an ancient, pre-evolutionary pedigree (see A. O. Lovejoy's classic, *The Great Chain of Being,* 1936). Consider, for example, Alexander Pope's *Essay on Man,* written early in the eighteenth century:

> Far as creation's ample range extends,
> The scale of sensual, mental powers ascends:
> Mark how it mounts, to man's imperial race,
> From the green myriads in the peopled grass.

And note a famous version from the very end of that century (figure 1.3). In his *Regular Gradation in Man,* British physician Charles White shoehorned all the ramifying diversity of vertebrate life into a single motley sequence running from birds through crocodiles and dogs, past apes, and up the conventional racist ladder of human groups to a Caucasian paragon, described with the rococo flourish of White's dying century:

> Where shall we find, unless in the European, that nobly arched head, con-

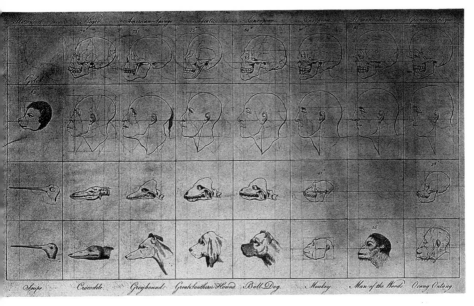

1.3. The linear gradations of the chain of being according to Charles White (1799). A motley sequence runs from birds to crocodiles to dogs and monkeys (bottom two rows), and then up the conventional racist ladder of human groups (top two rows).

taining such a quantity of brain . . . ? Where the perpendicular face, the prominent nose, and round projecting chin? Where that variety of features, and fullness of expression, . . . those rosy cheeks and coral lips? (White, 1799).

This tradition never vanished, even in our more enlightened age. In 1915, Henry Fairfield Osborn celebrated the linear accretion of cognition in a figure full of illuminating errors (figure 1.4). Chimps are not ancestors but modern cousins, equally distant in evolutionary terms from the unknown forebear of African great apes and humans. *Pithecanthropus* (*Homo erectus* in modern terms) is a potential ancestor, and the only legitimate member of the sequence. The inclusion of Piltdown is especially revealing. We now know that Piltdown was a fraud composed of a modern human cranium and an ape's jaw. As a contemporary cranium, Piltdown possessed a brain of modern size; yet so convinced were Osborn's colleagues that human fossils must show intermediate values on a ladder of progress, that they reconstructed Piltdown's brain according to their expectations. As for Neanderthal, these creatures were probably close cousins

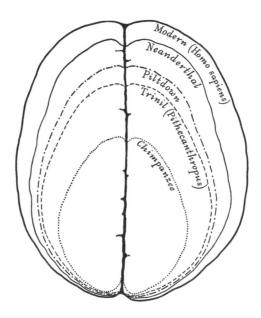

1.4. Progress in the evolution of the human brain as illustrated by Henry Fairfield Osborn in 1915.

1.5. A personally embarrassing illustration of our allegiance to the iconography of the march of progress. My books are dedicated to debunking this picture of evolution, but I have no control over jacket designs for foreign translations. Four translations of my books have used the "march of human progress" as a jacket illustration. This is from the Dutch translation of *Ever Since Darwin*.

belonging to a separate species, not ancestors. In any case, they had brains as large as ours, or larger, Osborn's ladder notwithstanding.

Nor have we abandoned this iconography in our generation. Consider figure 1.5, from a Dutch translation of one of my own books! The march of progress, single file, could not be more graphic. Lest we think that only Western culture promotes this conceit, I present one example of its spread (figure 1.6) purchased at the bazaar of Agra in 1985.

The march of progress is *the* canonical representation of evolution—the one picture immediately grasped and viscerally understood by all. This may best be appreciated by its prominent use in humor and in advertising. These professions provide our best test of public perceptions. Jokes and ads must click in the fleeting second that our attention grants them. Consider

1.6. I bought this children's science magazine in the bazaar of Agra, in India. The false iconography of the march of progress now has cross-cultural acceptance.

THE EVOLUTION OF MAN

1.7. A cartoonist can put the iconography of the ladder to good use. This example by Larry Johnson appeared in the *Boston Globe* before a Patriots–Raiders game.

figure 1.7, a cartoon drawn by Larry Johnson for the *Boston Globe* before a Patriots–Raiders football game. Or figure 1.8, by the cartoonist Szep, on the proper place of terrorism. Or figure 1.9, by Bill Day, on "scientific creationism." Or figure 1.10, by my friend Mike Peters, on the social possibilities traditionally open to men and to women. For advertising, consider the evolution of Guinness stout (figure 1.11) and of rental television (figure 1.12).*

The straitjacket of linear advance goes beyond iconography to the definition of evolution: the word itself becomes a synonym for *progress*. The makers of Doral cigarettes once presented a linear sequence of "improved" products through the years, under the heading "Doral's theory of evolution."† (Perhaps they are now embarrassed by this misguided claim, since

---

*Invoking another aspect of the same image—the equation of old and extinct with inadequate—Granada exhorts us to rent rather than buy because "today's latest models could be obsolete before you can say brontosaurus."

†Wonderfully ironic, since the sequence showed, basically, more effective filters. Evolution, to professionals, is adaptation to changing environments, not progress. Since the filters were responses to new conditions—public knowledge of health dangers—Doral did use the term *evolution* properly. Surely, however, they intended "absolutely better" rather than "punting to maintain profit"—a rather grisly claim in the light of several million deaths attributable to cigarette smoking.

A place in history

1.8. World terrorism parachutes into its appropriate place in the march of progress. By Szep, in the *Boston Globe*.

1.9. A "scientific creationist" takes his appropriate place in the march of progress. By Bill Day, in the *Detroit Free Press*.

1.10.   More mileage from the iconography of the ladder. By Mike Peters, in the *Dayton Daily News.* (Reprinted by permission of UFS, Inc.)

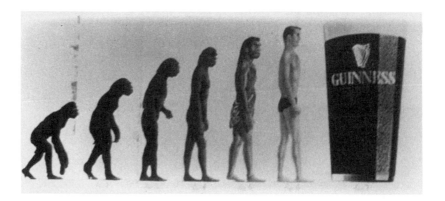

1.11.   The highest stage of human advance as photographed from an English billboard.

they refused me permission to reprint the ad.) Or consider an episode from the comic strip *Andy Capp* (figure 1.13). Flo has no problem in accepting evolution, but she defines it as progress, and views Andy's quadrupedal homecoming as quite the reverse.

Life is a copiously branching bush, continually pruned by the grim reaper of extinction, not a ladder of predictable progress. Most people may know this as a phrase to be uttered, but not as a concept brought into the deep interior of understanding. Hence we continually make errors inspired by unconscious allegiance to the ladder of progress, even when we explicitly deny such a superannuated view of life. For example, consider two errors, the second providing a key to our conventional misunderstanding of the Burgess Shale.

First, in an error that I call "life's little joke" (Gould, 1987a), we are virtually compelled to the stunning mistake of citing unsuccessful lineages as classic "textbook cases" of "evolution." We do this because we try to extract a single line of advance from the true topology of copious branching. In this misguided effort, we are inevitably drawn to bushes so near the brink of total annihilation that they retain only one surviving twig. We then view this twig as the acme of upward achievement, rather than the probable last gasp of a richer ancestry.

GRANADA TV RENTAL'S THEORY OF EVOLUTION.

1.12. The march of progress as portrayed in another advertisement.

1.13. The vernacular equation of evolution with progress. Andy's quadrupedal posture is interpreted as evolution in reverse. (By permission of © M.G.N. 1989, Syndication International/North America Syndicate, Inc.)

Consider the great warhorse of tradition—the evolutionary ladder of horses themselves (figure 1.14). To be sure, an unbroken evolutionary connection does link *Hyracotherium* (formerly called *Eohippus*) to modern *Equus*. And, yes again, modern horses are bigger, with fewer toes and higher crowned teeth. But *Hyracotherium–Equus* is not a ladder, or even a central lineage. This sequence is but one labyrinthine pathway among thousands on a complex bush. This particular route has achieved prominence for just one ironic reason—because all other twigs are extinct. *Equus* is the only twig left, and hence the tip of a ladder in our false iconography. Horses have become the classic example of progressive evolution because their bush has been so unsuccessful. We never grant proper acclaim to the real triumphs of mammalian evolution. Who ever hears a story about the evolution of bats, antelopes, or rodents—the current champions of mammalian life? We tell no such tales because we cannot linearize the bounteous success of these creatures into our favored ladder. They present us with thousands of twigs on a vigorous bush.

Need I remind everyone that at least one other lineage of mammals, especially dear to our hearts for parochial reasons, shares with horses both the topology of a bush with one surviving twig, and the false iconography of a march to progress?

In a second great error, we may abandon the ladder and acknowledge the branching character of evolutionary lineages, yet still portray the tree of life in a conventional manner chosen to validate our hopes for predictable progress.

The tree of life grows with a few crucial constraints upon its form. First, since any well-defined taxonomic group can trace its origin to a single

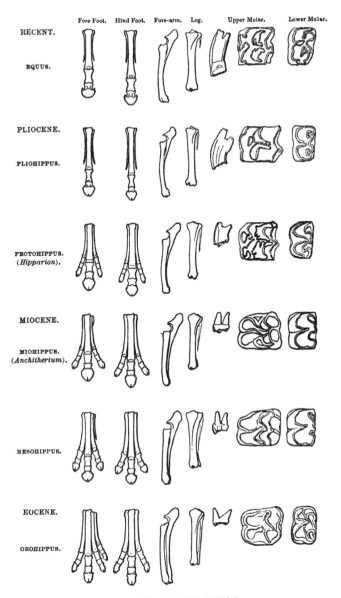

| Fore Foot. | Hind Foot. | Fore-arm. | Leg. | Upper Molar. | Lower Molar. |

RECENT.

EQUUS.

PLIOCENE.

PLIOHIPPUS.

PROTOHIPPUS.
(*Hipparion*).

MIOCENE.

MIOHIPPUS.
(*Anchitherium*).

MESOHIPPUS.

EOCENE.

OROHIPPUS.

GENEALOGY OF THE HORSE.

1.14.   The original version of the ladder of progress for horses, drawn by the
American paleontologist O. C. Marsh for Thomas Henry Huxley after Marsh
had shown his recently collected Western fossils to Huxley on his only visit to
the United States. Marsh convinced his English visitor about this sequence,
thus compelling Huxley to revamp his lecture on the evolution of horses given
in New York in 1876. Note the steady decrease in number of toes and
increase in height of teeth. Since Marsh drew all his specimens the same size,
we do not see the other classical trend of increase in stature.

common ancestor, an evolutionary tree must have a unique basal trunk.*
Second, all branches of the tree either die or ramify further. Separation is
irrevocable; distinct branches do not join.†

Yet, within these constraints of *monophyly* and *divergence,* the geometric possibilities for evolutionary trees are nearly endless. A bush may quickly expand to maximal width and then taper continuously, like a Christmas tree. Or it may diversify rapidly, but then maintain its full width by a continuing balance of innovation and death. Or it may, like a tumbleweed, branch helter-skelter in a confusing jumble of shapes and sizes.

Ignoring these multifarious possibilities, conventional iconography has fastened upon a primary model, the "cone of increasing diversity," an upside-down Christmas tree. Life begins with the restricted and simple, and progresses ever upward to more and more and, by implication, better and better. Figure 1.15 on the evolution of coelomates (animals with a body cavity, the subjects of this book), shows the orderly origin of everything from a simple flatworm. The stem splits to a few basic stocks; none becomes extinct; and each diversifies further, into a continually increasing number of subgroups.

*A properly defined group with a single common ancestor is called monophyletic. Taxonomists insist upon monophyly in formal classification. However, many vernacular names do not correspond to well-constituted evolutionary groups because they include creatures with disparate ancestries—"polyphyletic" groups in technical parlance. For example, folk classifications that include bats among birds, or whales among fishes, are polyphyletic. The vernacular term *animal* itself probably denotes a polyphyletic group, since sponges (almost surely), and probably corals and their allies as well, arose separately from unicellular ancestors—while all other animals of our ordinary definitions belong to a third distinct group. The Burgess Shale contains numerous sponges, and probably some members of the coral phylum as well, but this book will treat only the third great group—the coelomates, or animals with a body cavity. The coelomates include all vertebrates and all common invertebrates except sponges, corals, and their allies. Since the coelomates are clearly monophyletic (Hanson, 1977), the subjects of this book form a proper evolutionary group.

†This fundamental principle, while true for the complex multicellular animals treated in this book, does not apply to all life. Hybridization between distant lineages occurs frequently in plants, producing a "tree of life" that often looks more like a network than a conventional bush. (I find it amusing that the classic metaphor of the tree of life, used as a picture of evolution ever since Darwin and so beautifully accurate for animals, may not apply well to plants, the source of the image.) In addition, we now know that genes can be transferred laterally, usually by viruses, across species boundaries. This process may be important in the evolution of some unicellular creatures, but probably plays only a small role in the phylogeny of complex animals, if only because two embryological systems based upon intricately different developmental pathways cannot mesh, films about flies and humans notwithstanding.

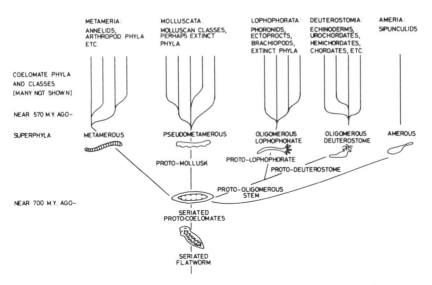

1.15. A recent iconography for the evolution of coelomate animals, drawn according to the convention of the cone of increasing diversity (Valentine, 1977).

Figure 1.16 presents a panoply of cones drawn from popular modern textbooks—three abstract and three actual examples for groups crucial to the argument of this book. (In chapter IV, I discuss the origin of this model in Haeckel's original trees and their influence upon Walcott's great error in reconstructing the Burgess fauna.) All these trees show the same pattern: branches grow ever upward and outward, splitting from time to time. If some early lineages die, later gains soon overbalance these losses. Early deaths can eliminate only small branches near the central trunk. Evolution unfolds as though the tree were growing up a funnel, always filling the continually expanding cone of possibilities..

In its conventional interpretation, the cone of diversity propagates an interesting conflation of meanings. The horizontal dimension shows diversity—fishes plus insects plus snails plus starfishes at the top take up much more lateral room than just flatworms at the bottom. But what does the vertical dimension represent? In a literal reading, up and down should record only younger and older in geological time: organisms at the neck of the funnel are ancient; those at the lip, recent. But we also read upward movement as simple to complex, or primitive to advanced. *Placement in time is conflated with judgment of worth.*

Our ordinary discourse about animals follows this iconography. Nature's

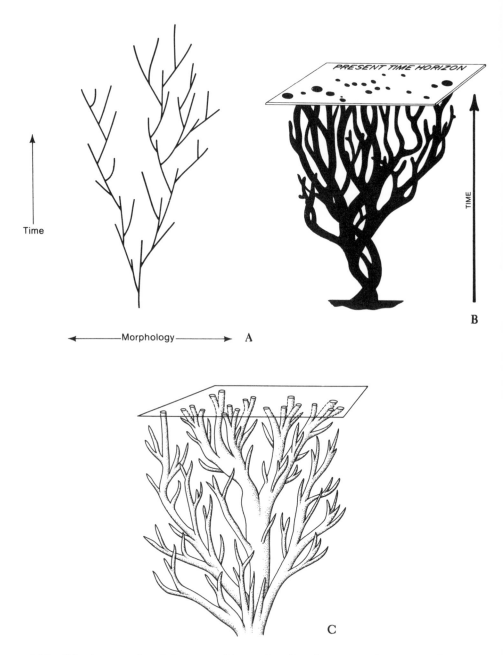

Time

Morphology   A

PRESENT TIME HORIZON

TIME

B

C

1.16. The iconography of the cone of increasing diversity, as seen in six examples from textbooks. All these diagrams are presented as simple objective portrayals of evolution; none are explicit representations of diversification as opposed to some other evolutionary process. Three abstract examples (A–C) are followed by conventional views of three specific phylogenies–vertebrate (D), arthropod (E), and mammalian (F, on p. 42). The data of the Burgess Shale falsify this central view of arthropod evolution as a continuous process of increasing diversification.

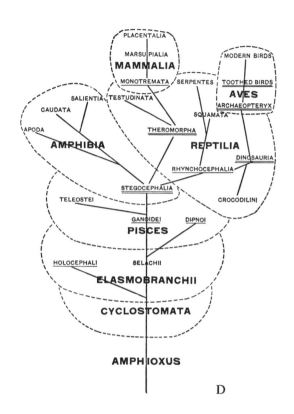

D

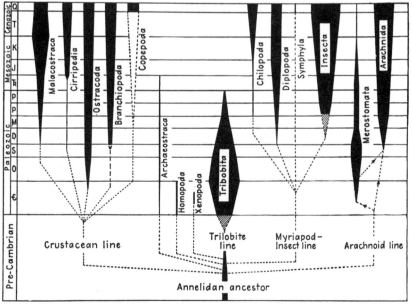

E

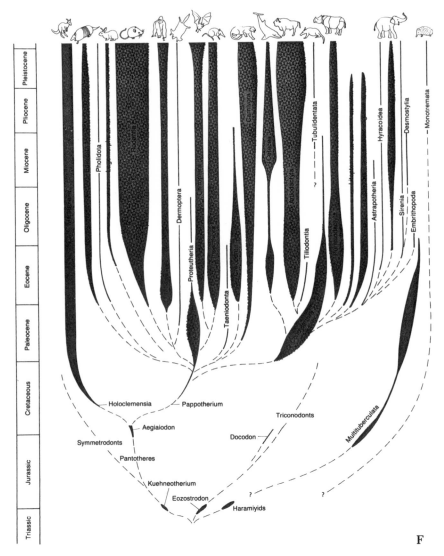

1.16 (*continued*).   A conventional view of mammalian phylogeny.

theme is diversity. We live surrounded by coeval twigs of life's tree. In Darwin's world, all (as survivors in a tough game) have some claim to equal status. Why, then, do we usually choose to construct a ranking of implied worth (by assumed complexity, or relative nearness to humans, for example)? In a review of a book on courtship in the animal kingdom, Jonathan Weiner (*New York Times Book Review*, March 27, 1988) describes the author's scheme of organization: "Working in loosely evolutionary order, Mr. Walters begins with horseshoe crabs, which have been meeting and mating on dark beaches in synchrony with tide and moon for 200 million years." Later chapters make the "long evolutionary leap to the antics of the pygmy chimpanzee." Why is this sequence called "evolutionary order"? Anatomically complex horseshoe crabs are not ancestral to vertebrates; the two phyla, Arthropoda and Chordata, have been separate from the very first records of multicellular life.

In another recent example, showing that this error infests technical as well as lay discourse, an editorial in *Science,* the leading scientific journal in America, constructs an order every bit as motley and senseless as White's "regular gradation" (see figure 1.3). Commenting on species commonly used for laboratory work, the editors discuss the "middle range" between unicellular creatures and guess who at the apex: "Higher on the evolutionary ladder," we learn, "the nematode, the fly and the frog have the advantage of complexity beyond the single cell, but represent far simpler species than mammals" (June 10, 1988).

The fatuous idea of a single order amidst the multifarious diversity of modern life flows from our conventional iconographies and the prejudices that nurture them—the ladder of life and the cone of increasing diversity. By the ladder, horseshoe crabs are judged as simple; by the cone, they are deemed old.* And one implies the other under the grand conflation discussed above—down on the ladder also means old, while low on the cone denotes simple.

I don't think that any particular secret, mystery, or inordinate subtlety underlies the reasons for our allegiance to these false iconographies of ladder and cone. They are adopted because they nurture our hopes for a universe of intrinsic meaning defined in our terms. We simply cannot bear the implications of Omar Khayyám's honesty:

> Into this Universe, and Why not knowing,
> Nor whence, like Water willy-nilly flowing:
>   And out of it, as Wind along the Waste
> I know not Whither, willy-nilly blowing.

A later quatrain of the *Rubáiyát* proposes a counteracting strategy, but acknowledges its status as a vain hope:

---

*Another factual irony: despite the usual picture of horseshoe crabs as "living fossils," *Limulus polyphemus* (our American East Coast species) has no fossil record whatever. The genus *Limulus* ranges back only some 20 million years, not 200 million. We mistakenly regard horseshoe crabs as "living fossils" because the group has never produced many species, and therefore never developed much evolutionary potential for diversification; consequently, modern species are morphologically similar to early forms. But the species themselves are not notably old.

> Ah Love! could you and I with Fate conspire
> To grasp this sorry Scheme of Things entire,
>   Would we not shatter it to bits—and then
> Re-mold it nearer to the Heart's Desire!

Most myths and early scientific explanations of Western culture pay homage to this "heart's desire." Consider the primal tale of Genesis, presenting a world but a few thousand years old, inhabited by humans for all but the first five days, and populated by creatures made for our benefit and subordinate to our needs. Such a geological background could inspire Alexander Pope's confidence, in the *Essay on Man,* about the deeper meaning of immediate appearances:

> All Nature is but art, unknown to thee;
> All chance, direction, which thou canst not see;
> All discord, harmony not understood;
> All partial evil, universal good.

But, as Freud observed, our relationship with science must be paradoxical because we are forced to pay an almost intolerable price for each major gain in knowledge and power—the psychological cost of progressive dethronement from the center of things, and increasing marginality in an uncaring universe. Thus, physics and astronomy relegated our world to a corner of the cosmos, and biology shifted our status from a simulacrum of God to a naked, upright ape.

To this cosmic redefinition, my profession contributed its own special shock—geology's most frightening fact, we might say. By the turn of the last century, we knew that the earth had endured for millions of years, and that human existence occupied but the last geological millimicrosecond of this history—the last inch of the cosmic mile, or the last second of the geological year, in our standard pedagogical metaphors.

We cannot bear the central implication of this brave new world. If humanity arose just yesterday as a small twig on one branch of a flourishing tree, then life may not, in any genuine sense, exist for us or because of us. Perhaps we are only an afterthought, a kind of cosmic accident, just one bauble on the Christmas tree of evolution.

What options are left in the face of geology's most frightening fact? Only two, really. We may, as this book advocates, accept the implications and learn to seek the meaning of human life, including the source of morality, in other, more appropriate, domains—either stoically with a sense of loss, or with joy in the challenge if our temperament be optimistic. Or we may continue to seek cosmic comfort in nature by reading life's history in a distorted light.

If we elect the second strategy, our maneuvers are severely restricted by our geological history. When we infested all but the first five days of time, the history of life could easily be rendered in our terms. But if we wish to assert human centrality in a world that functioned without us until the last moment, we must somehow grasp all that came before as a grand preparation, a foreshadowing of our eventual origin.

The old chain of being would provide the greatest comfort, but we now know that the vast majority of "simpler" creatures are not human ancestors or even prototypes, but only collateral branches on life's tree. The cone of increasing progress and diversity therefore becomes our iconography of choice. The cone implies predictable development from simple to complex, from less to more. *Homo sapiens* may form only a twig, but if life moves, even fitfully, toward greater complexity and higher mental powers, then the eventual origin of self-conscious intelligence may be implicit in all that came before. In short, I cannot understand our continued allegiance to the manifestly false iconographies of ladder and cone except as a desperate finger in the dike of cosmically justified hope and arrogance.

I leave the last word on this subject to Mark Twain, who grasped so graphically, when the Eiffel Tower was the world's tallest building, the implications of geology's most frightening fact:

> Man has been here 32,000 years. That it took a hundred million years to prepare the world for him* is proof that that is what it was done for. I suppose it is. I dunno. If the Eiffel Tower were now representing the world's age, the skin of paint on the pinnacle knob at its summit would represent man's share of that age; and anybody would perceive that the skin was what the tower was built for. I reckon they would, I dunno.

REPLAYING LIFE'S TAPE: THE CRUCIAL EXPERIMENT

The iconography of the cone made Walcott's original interpretation of the Burgess fauna inevitable. Animals so close in time to the origin

---

*Twain used Lord Kelvin's estimate, then current, for the age of the earth. The estimated ages have lengthened substantially since then, but Twain's proportions are not far off. He took human existence as about 1/3000 of the earth's age. At current estimates of 250,000 years for the origin of our species, *Homo sapiens*, the earth would be 0.75 billion years old if our span were 1/3000 of totality. By best current estimates, the earth is 4.5 billion years old.

The Cone of Increasing Diversity

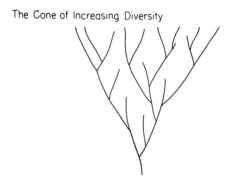

Decimation and Diversification

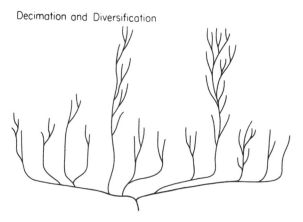

1.17. The false but still conventional iconography of the cone of increasing diversity, and the revised model of diversification and decimation, suggested by the proper reconstruction of the Burgess Shale.

of multicellular life would have to lie in the narrow neck of the funnel. Burgess animals therefore could not stray beyond a strictly limited diversity and a basic anatomical simplicity. In short, they had to be classified either as primitive forms within modern groups, or as ancestral animals that might, with increased complexity, progress to some familiar form of the modern seas. Small wonder, then, that Walcott interpreted every organism in the Burgess Shale as a primitive member of a prominent branch on life's later tree.

I know no greater challenge to the iconography of the cone—and hence no more important case for a fundamentally revised view of life—than the radical reconstructions of Burgess anatomy presented by Whittington and his colleagues. They have literally followed our most venerable metaphor for revolution: they have turned the traditional interpretation on its head. By recognizing so many unique anatomies in the Burgess, and by showing that familiar groups were then experimenting with designs so far beyond the modern range, they have inverted the cone. The sweep of anatomical variety reached a maximum right after the initial diversification of multi-cellular animals. The later history of life proceeded by elimination, not expansion. The current earth may hold more species than ever before, but most are iterations upon a few basic anatomical designs. (Taxonomists have described more than a half million species of beetles, but nearly all are minimally altered Xeroxes of a single ground plan.) In fact, the probable increase in number of species through time merely underscores the puzzle and paradox. Compared with the Burgess seas, today's oceans contain many more species based upon many fewer anatomical plans.

Figure 1.17 presents a revised iconography reflecting the lessons of the Burgess Shale. The maximum range of anatomical possibilities arises with the first rush of diversification. Later history is a tale of restriction, as most of these early experiments succumb and life settles down to generating endless variants upon a few surviving models.*

---

*I have struggled over a proper name for this phenomenon of massive elimination from an initial set of forms, with concentration of all future history into a few surviving lineages. For many years, I thought of this pattern as "winnowing," but must now reject this metaphor because all meanings of winnowing refer to separation of the good from the bad (grain from chaff in the original)—while I believe that the preservation of only a few Burgess possibilities worked more like a lottery.

I have finally decided to describe this pattern as "decimation," because I can combine the literal and vernacular senses of this word to suggest the two cardinal aspects stressed throughout this book: the largely random sources of survival or death, and the high overall probability of extinction.

Randomness. "Decimate" comes from the Latin *decimare*, "to take one in ten." The word refers to a standard punishment applied in the Roman army to groups of soldiers guilty of mutiny, cowardice, or some other crime. One soldier of every ten was selected by lot and put to death. I could not ask for a better metaphor of extinction by lottery.

Magnitude. But the literal meaning might suggest the false implication that chances for death, though applied equally to all, are rather low—only about 10 percent. The Burgess pattern indicates quite the opposite. Most die and few are chosen—a 90 percent chance of death would be a good estimate for major Burgess lineages. In modern vernacular English, "decimate" has come to mean "destroy an overwhelming majority," rather than the small percentage of the ancient Roman practice. The *Oxford English Dictionary* indicates that this revised usage is not an error or a reversed meaning, but has its own pedigree—for "decimation" has also been used for the taking of nine in ten.

In any case, I wish to join the meaning of randomness explicit in the original Roman definition with the modern implication that most die and only a few survive. In this combined sense, decimation is the right metaphor for the fate of the Burgess Shale fauna—random elimination of most lineages.

This inverted iconography, however interesting and radical in itself, need not imply a revised view of evolutionary predictability and direction. We can abandon the cone, and accept the inverted iconography, yet still maintain full allegiance to tradition if we adopt the following interpretation: all but a small percentage of Burgess possibilities succumbed, but the losers were chaff, and predictably doomed. Survivors won for cause—and cause includes a crucial edge in anatomical complexity and competitive ability.

But the Burgess pattern of elimination also suggests a truly radical alternative, precluded by the iconography of the cone. Suppose that winners have not prevailed for cause in the usual sense. Perhaps the grim reaper of anatomical designs is only Lady Luck in disguise. Or perhaps the actual reasons for survival do not support conventional ideas of cause as complexity, improvement, or anything moving at all humanward. Perhaps the grim reaper works during brief episodes of mass extinction, provoked by unpredictable environmental catastrophes (often triggered by impacts of extraterrestrial bodies). Groups may prevail or die for reasons that bear no relationship to the Darwinian basis of success in normal times. Even if fishes hone their adaptations to peaks of aquatic perfection, they will all die if the ponds dry up. But grubby old Buster the Lungfish, former laughingstock of the piscine priesthood, may pull through—and not because a bunion on his great-grandfather's fin warned his ancestors about an impending comet. Buster and his kin may prevail because a feature evolved long ago for a different use has fortuitously permitted survival during a sudden and unpredictable change in rules. And if we are Buster's legacy, and the result of a thousand other similarly happy accidents, how can we possibly view our mentality as inevitable, or even probable?

We live, as our humorists proclaim, in a world of good news and bad news. The good news is that we can specify an experiment to decide between the conventional and the radical interpretations of extinction, thereby settling the most important question we can ask about the history of life. The bad news is that we can't possibly perform the experiment.

I call this experiment "replaying life's tape." You press the rewind button and, making sure you thoroughly erase everything that actually happened, go back to any time and place in the past—say, to the seas of the Burgess Shale. Then let the tape run again and see if the repetition looks at all like the original. If each replay strongly resembles life's actual pathway, then we must conclude that what really happened pretty much had to occur. But suppose that the experimental versions all yield sensible results strikingly different from the actual history of life? What could we then say about the predictability of self-conscious intelligence? or of mammals? or

THE MEANINGS OF DIVERSITY AND DISPARITY

I must introduce at this point an important distinction that should allay a classic source of confusion. Biologists use the vernacular term diversity in several different technical senses. They may talk about "diversity" as number of distinct species in a group: among mammals, rodent diversity is high, more than 1,500 separate species; horse diversity is low, since zebras, donkeys, and true horses come in fewer than ten species. But biologists also speak of "diversity" as difference in body plans. Three blind mice of differing species do not make a diverse fauna, but an elephant, a tree, and an ant do—even though each assemblage contains just three species.

The revision of the Burgess Shale rests upon its diversity in this second sense of disparity in anatomical plans. Measured as number of species, Burgess diversity is not high. This fact embodies a central paradox of early life: How could so much disparity in body plans evolve in the apparent absence of substantial diversity in number of species?—for the two are correlated, more or less in lockstep, by the iconography of the cone (see figure 1.16).

When I speak of decimation, I refer to reduction in the number of anatomical designs for life, not numbers of species. Most paleontologists agree that the simple count of species has augmented through time (Sepkoski et al., 1981)—and this increase of species must therefore have occurred within a reduced number of body plans.

Most people do not fully appreciate the stereotyped character of current life. We learn lists of odd phyla in high school, until kinorhynch, priapulid, gnathostomulid, and pogonophoran roll off the tongue (at least until the examination ends). Focusing on a few oddballs, we forget how unbalanced life can be. Nearly 80 percent of all described animal species are arthropods (mostly insects). On the sea floor, once you enumerate polychaete worms, sea urchins, crabs, and snails, there aren't that many coelomate invertebrates left. Stereotypy, or the cramming of most species into a few anatomical plans, is a cardinal feature of modern life—and its greatest difference from the world of Burgess times.

Several of my colleagues (Jaanusson, 1981; Runnegar, 1987) have suggested that we eliminate the confusion about diversity by restricting this vernacular term to the first sense—number of species. The second sense—difference in body plans—should then be called disparity. Using this terminology, we may acknowledge a central and surprising fact of life's history—marked decrease in disparity followed by an outstanding increase in diversity within the few surviving designs.

of vertebrates? or of life on land? or simply of multicellular persistence for 600 million difficult years?

We can now appreciate the central importance of the Burgess revision and its iconography of decimation. With the ladder or the cone, the issue of life's tape does not arise. The ladder has but one bottom rung, and one direction. Replay the tape forever, and *Eohippus* will always gallop into the sunrise, bearing its ever larger body on fewer toes. Similarly, the cone has a narrow neck and a restricted range of upward movement. Rewind the tape back into the neck of time, and you will always obtain the same prototypes, constrained to rise in the same general direction.

But if a radical decimation of a much greater range of initial possibilities determined the pattern of later life, including the chance of our own origin, then consider the alternatives. Suppose that ten of a hundred designs will survive and diversify. If the ten survivors are predictable by superiority of anatomy (interpretation 1), then they will win each time— and Burgess eliminations do not challenge our comforting view of life. But if the ten survivors are protégés of Lady Luck or fortunate beneficiaries of odd historical contingencies (interpretation 2), then each replay of the tape will yield a different set of survivors and a radically different history. And if you recall from high-school algebra how to calculate permutations and combinations, you will realize that the total number of combinations for 10 items from a pool of 100 yields more than 17 trillion potential outcomes. I am willing to grant that some groups may have enjoyed an edge (though we have no idea how to identify or define them), but I suspect that the second interpretation grasps a central truth about evolution. The Burgess Shale, in making this second interpretation intelligible by the hypothetical experiment of the tape, promotes a radical view of evolutionary pathways and predictability.

Rejection of ladder and cone does not throw us into the arms of a supposed opposite—pure chance in the sense of coin tossing or of God playing dice with the universe. Just as the ladder and the cone are limiting iconographies for life's history, so too does the very idea of dichotomy grievously restrict our thinking. Dichotomy has its own unfortunate iconography—a single line embracing all possible opinions, with the two ends representing polar opposites—in this case, determinism and randomness.

An old tradition, dating at least to Aristotle, advises the prudent person to stake out a position comfortably toward the middle of the line—the *aurea mediocritas* ("golden mean"). But in this case the middle of the line has not been so happy a place, and the game of dichotomy has seriously hampered our thinking about the history of life. We may understand that the older determinism of predictable progress cannot strictly apply, but we think that our only alternative lies with the despair of pure randomness. So we are driven back toward the old view, and finish, with discomfort, at some ill-defined confusion in between.

I strongly reject any conceptual scheme that places our options on a line, and holds that the only alternative to a pair of extreme positions lies somewhere between them. More fruitful perspectives often require that we step off the line to a site outside the dichotomy.

I write this book to suggest a third alternative, off the line. I believe that the reconstructed Burgess fauna, interpreted by the theme of replaying life's tape, offers powerful support for this different view of life: any replay of the tape would lead evolution down a pathway radically different from the road actually taken. But the consequent differences in outcome do not imply that evolution is senseless, and without meaningful pattern; the divergent route of the replay would be just as interpretable, just as explainable *after* the fact, as the actual road. But the diversity of possible itineraries does demonstrate that eventual results cannot be predicted at the outset. Each step proceeds for cause, but no finale can be specified at the start, and none would ever occur a second time in the same way, because any pathway proceeds through thousands of improbable stages. Alter any early event, ever so slightly and without apparent importance at the time, and evolution cascades into a radically different channel.

This third alternative represents no more nor less than the essence of history. Its name is contingency—and contingency is a thing unto itself, not the titration of determinism by randomness. Science has been slow to admit the different explanatory world of history into its domain—and our interpretations have been impoverished by this omission. Science has also tended to denigrate history, when forced to a confrontation, by regarding any invocation of contingency as less elegant or less meaningful than explanations based directly on timeless "laws of nature."

This book is about the nature of history and the overwhelming improbability of human evolution under themes of contingency and the metaphor of replaying life's tape. It focuses upon the new interpretation of the Burgess Shale as our finest illustration of what contingency implies in our quest to understand the evolution of life.

I concentrate upon details of the Burgess Shale because I don't believe that important concepts should be discussed tendentiously in the abstract (much as I have disobeyed the rule in this opening chapter!). People, as curious primates, dote on concrete objects that can be seen and fondled. God dwells among the details, not in the realm of pure generality. We must tackle and grasp the larger, encompassing themes of our universe, but we make our best approach through small curiosities that rivet our attention—all those pretty pebbles on the shoreline of knowledge. For the ocean of truth washes over the pebbles with every wave, and they rattle and clink with the most wondrous din.

We can argue about abstract ideas forever. We can posture and feint. We can "prove" to the satisfaction of one generation, only to become the laughingstock of a later century (or, worse still, to be utterly forgotten). We may even validate an idea by grafting it permanently upon an object of nature—thus participating in the legitimate sense of a great human adventure called "progress in scientific thought."

But the animals of the Burgess Shale are somehow even more satisfying in their adamantine factuality. We will argue forever about the meaning of life, but *Opabinia* either did or did not have five eyes—and we can know for certain one way or the other. The animals of the Burgess Shale are also the world's most important fossils, in part because they have revised our view of life, but also because they are objects of such exquisite beauty. Their loveliness lies as much in the breadth of ideas that they embody, and in the magnitude of our struggle to interpret their anatomy, as in their elegance of form and preservation.

The animals of the Burgess Shale are holy objects—in the unconventional sense that this word conveys in some cultures. We do not place them on pedestals and worship from afar. We climb mountains and dynamite hillsides to find them. We quarry them, split them, carve them, draw them, and dissect them, struggling to wrest their secrets. We vilify and curse them for their damnable intransigence. They are grubby little creatures of a sea floor 530 million years old, but we greet them with awe because they are the Old Ones, and they are trying to tell us something.

CHAPTER II

A Background for the Burgess Shale

LIFE BEFORE THE BURGESS: THE CAMBRIAN EXPLOSION AND THE ORIGIN OF ANIMALS

Soured, perhaps, by memories of the multiplication tables, college students hate the annual ritual of memorizing the geological time scale in introductory courses on the history of life. We professors insist, claiming this venerable sequence as our alphabet. The entries are cumbersome— Cambrian, Ordovician, Silurian—and refer to such arcana as Roman names for Wales and threefold divisions of strata in Germany. We use little tricks and enticements to encourage compliance. For years, I held a mnemonics contest for the best entry to replace the traditional and insipid "Campbell's ordinary soup does make Peter pale . . ." or the underground salacious versions that I would blush to record, even here. During political upheavals of the early seventies, my winner (for epochs of the Tertiary, see figure 2.1) read: "Proletarian efforts off many pig police. Right on!" The all-time champion reviewed a porno movie called *Cheap Meat*—with perfect rhyme and scansion and only one necessary neologism, easily interpreted, at the end of the third line. This entry proceeds in unconventional order, from latest to earliest, and lists all the eras first, then all the periods:

Cheap Meat performs passably,
Quenching the celibate's jejune thirst,
Portraiture, presented massably,
Drowning sorrow, oneness cursed.

The winner also provided an epilogue, for the epochs of the Cenozoic era:

Rare pornography, purchased meekly
O Erogeny, Paleobscene.*

When such blandishments fail, I always say, try an honest intellectual argument: if these names were arbitrary divisions in a smooth continuum of events unfolding through time, I would have some sympathy for the opposition—for then we might take the history of modern multicellular life, about 600 million years, and divide this time into even and arbitrary units easily remembered as 1–12 or A–L, at 50 million years per unit.

But the earth scorns our simplifications, and becomes much more interesting in its derision. The history of life is not a continuum of development, but a record punctuated by brief, sometimes geologically instantaneous, episodes of mass extinction and subsequent diversification. The geological time scale maps this history, for fossils provide our chief criterion in fixing the temporal order of rocks. The divisions of the time scale are set at these major punctuations because extinctions and rapid diversifications leave such clear signatures in the fossil record. Hence, the time scale is not a devil's ploy for torturing students, but a chronicle of key moments in life's history. By memorizing those infernal names, you learn the major episodes of earthly time. I make no apologies for the central importance of such knowledge.

The geological time scale (figure 2.1) is divided hierarchically into eras, periods, and epochs. The boundaries of the largest divisions—the eras—mark the greatest events. Of the three era boundaries, two designate the most celebrated of mass extinctions. The late Cretaceous mass extinction, some 65 million years ago, sets the boundary between Mesozoic and Cenozoic eras. Although not the largest of "great dyings," this event surpasses all others in fame, for dinosaurs perished in its wake, and the evolution of large mammals (including, much later, ourselves) became possible as a result. The second boundary, between the Paleozoic and Mesozoic eras (225 million years ago), records the granddaddy of all extinctions—the late

---

*There are two in jokes in this line: *orogeny* is standard geological jargon for mountain building; *Paleobscene* is awfully close to the epoch's actual name—Paleocene.

Permian event that irrevocably set the pattern of all later history by extirpating up to 96 percent of marine species.

The third and oldest boundary, between Precambrian times and the Paleozoic era (about 570 million years ago), marks a different and more puzzling kind of event. A mass extinction may have occurred at or near this boundary, but the inception of the Paleozoic era denotes a concentrated episode of diversification—the "Cambrian explosion," or first appearance of multicellular animals with hard parts in the fossil record. The importance of the Burgess Shale rests upon its relationship to this pivotal moment in the history of life. The Burgess fauna does not lie within the explosion itself, but marks a time soon afterward, about 530 million years ago, before the relentless motor of extinction had done much work, and when the full panoply of results therefore stood on display. As the only

GEOLOGIC ERAS			
Era	Period	Epoch	Approximate number of years ago (millions of years)
Cenozoic	Quaternary	Holocene (Recent) Pleistocene	
	Tertiary	Pliocene Miocene Oligocene Eocene Paleocene	
Mesozoic	Cretaceous Jurassic Triassic		65
Paleozoic	Permian Carboniferous (Pennsylvanian and Mississippian) Devonian Silurian Ordovician Cambrian		225
Precambrian			570

2.1. The geological time scale.

major soft-bodied fauna from this primordial time, the Burgess Shale provides our sole vista upon the inception of modern life in all its fullness.

The Cambrian explosion is a tolerably ancient event, but the earth is 4.5 billion years old, so multicellular life of modern design occupies little more than 10 percent of earthly time. This chronology poses the two classic puzzles of the Cambrian explosion—enigmas that obsessed Darwin (1859, pp. 306–10) and remain central riddles of life's history: (1) Why did multicellular life appear so late? (2) And why do these anatomically complex creatures have no direct, simpler precursors in the fossil record of Precambrian times?

These questions are difficult enough now, in the context of a rich record of Precambrian life, all discovered since the 1950s. But when Charles Doolittle Walcott found the Burgess Shale in 1909, they seemed well-nigh intractable. In Walcott's time, the slate of Precambrian life was absolutely blank. Not a single well-documented fossil had been found from any time before the Cambrian explosion, and the earliest evidence of multicellular animals coincided with the earliest evidence of any life at all! From time to time, claims had been advanced—more than once by Walcott himself—for Precambrian animals, but none had withstood later scrutiny. These creatures of imagination had been founded upon hope, and were later exposed as ripple marks, inorganic precipitates, or genuine fossils of later epochs misdiagnosed as primordial.

This apparent absence of life during most of the earth's history, and its subsequent appearance at full complexity, posed no problem for anti-evolutionists. Roderick Impey Murchison, the great geologist who first worked out the record of early life, simply viewed the Cambrian explosion as God's moment of creation, and read the complexity of the first animals as a sign that God had invested appropriate care in his initial models. Murchison, writing five years before Darwin's *Origin of Species,* explicitly identified the Cambrian explosion as a disproof of evolution ("transmutation" in his terms), while he extolled the compound eye of the first trilobites as a marvel of exquisite design:

> The earliest signs of living things, announcing as they do a high complexity of organization, entirely exclude the hypothesis of a transmutation from lower to higher grades of being. The first fiat of Creation which went forth, doubtlessly ensured the perfect adaptation of animals to the surrounding media; and thus, whilst the geologist recognizes a beginning, he can see in the innumerable facets of the eye of the earliest crustacean, the same evidences of Omniscience as in the completion of the vertebrate form (1854, p. 459).

Darwin, honest as always in exposing the difficulties of his theory, placed the Cambrian explosion at the pinnacle of his distress, and devoted an entire section to this subject in the *Origin of Species*. Darwin acknowledged the anti-evolutionary interpretation of many important geologists: "Several of the most eminent geologists, with Sir. R. Murchison at their head, are convinced that we see in the organic remains of the lowest Silurian* stratum the dawn of life on this planet" (1859, p. 307). Darwin recognized that his theory required a rich Precambrian record of precursors for the first complex animals:

> If my theory be true, it is indisputable that before the lowest Silurian stratum was deposited, long periods elapsed, as long as, or probably far longer than, the whole interval from the Silurian age to the present day; and that during these vast, yet quite unknown periods of time, the world swarmed with living creatures (1859, p. 307).

Darwin invoked his standard argument to resolve this uncomfortable problem: the fossil record is so imperfect that we do not have evidence for most events of life's history. But even Darwin acknowledged that his favorite ploy was wearing a bit thin in this case. His argument could easily account for a missing stage in a single lineage, but could the agencies of imperfection really obliterate absolutely all evidence for positively every creature during most of life's history? Darwin admitted: "The case at present must remain inexplicable; and may be truly urged as a valid argument against the views here entertained" (1859, p. 308).

Darwin has been vindicated by a rich Precambrian record, all discovered in the past thirty years. Yet the peculiar character of this evidence has not matched Darwin's prediction of a continuous rise in complexity toward Cambrian life, and the problem of the Cambrian explosion has remained as stubborn as ever—if not more so, since our confusion now rests on knowledge, rather than ignorance, about the nature of Precambrian life.

Our Precambrian record now stretches back to the earliest rocks that could contain life. The earth is 4.5 billion years old, but heat from impacting bodies (as the planets first coalesced), and from radioactive decay of short-lived isotopes, caused our planet to melt and differentiate early in its history. The oldest sedimentary rocks—the 3.75-billion-year-old Isua series of west Greenland—record the cooling and stabilization of the earth's crust. These strata are too metamorphosed (altered by heat and pressure) to preserve the morphological remains of living creatures, but Schidlowski (1988) has recently argued that this oldest potential source of evidence

---

*The "lowest Silurian" refers to rocks now called Cambrian, a period not yet codified and accepted by all in 1859. Darwin is discussing the Cambrian explosion in this passage.

retains a chemical signature of organic activity. Of the two common isotopes of carbon, ^{12}C and ^{13}C, photosynthesis differentially uses the lighter ^{12}C and therefore raises the ratio of isotopes—$^{12}C/^{13}C$—above the values that would be measured if all the sedimentary carbon had an inorganic source. The Isua rocks show the enhanced values of ^{12}C that arise as a product of organic activity.*

Just as chemical evidence for life may appear in the first rocks capable of providing it, morphological remains are also as old as they could possibly be. Both stromatolites (mats of sediment trapped and bound by bacteria and blue-green algae) and actual cells have been found in the earth's oldest unmetamorphosed sediments, dating to 3.5–3.6 billion years in Africa and Australia (Knoll and Barghoorn, 1977; Walter, 1983).

Such a simple beginning would have pleased Darwin, but the later history of Precambrian life stands strongly against his assumption of a long and gradual rise in complexity toward the products of the Cambrian explosion. For 2.4 billion years after the Isua sediments, or nearly two-thirds of the entire history of life on earth, all organisms were single-celled creatures of the simplest, or prokaryotic, design. (Prokaryotic cells have no organelles—no nucleus, no paired chromosomes, no mitochondria, no chloroplasts. The much larger eukaryotic cells of other unicellular organisms, and of all multicellular creatures, are vastly more complex and may have evolved from colonies of prokaryotes; mitochondria and chloroplasts, at least, look remarkably like entire prokaryotic organisms and retain some DNA of their own, perhaps as a vestige of this former independence. Bacteria and blue-green algae, or cyanophytes, are prokaryotes. All other common unicellular organisms—including the *Amoeba* and *Paramecium* of high-school biology labs—are eukaryotes.)

The advent of eukaryotic cells in the fossil record some 1.4 billion years ago marks a major increment in life's complexity, but multicellular animals did not follow triumphantly in their wake. The time between the appearance of the first eukaryotic cell and the first multicellular animal is longer than the entire period of multicellular success since the Cambrian explosion.

The Precambrian record does contain one fauna of multicellular animals preceding the Cambrian explosion, the Ediacara fauna, named for a locality in Australia but now known from rocks throughout the world. But this

---

*Although the $^{12}C/^{13}C$ ratio in the Isua rocks is indicative of organic fractionation, the excess of ^{12}C is not so high as for later sediments. Schidlowski argues that the subsequent metamorphism of the Isua rocks lowered the ratio (while leaving it within the range of organic values), and that the original ratio probably matched that of later sediments.

fauna can offer no comfort to Darwin's expectation for two reasons. First, the Ediacara is barely Precambrian in age. These animals are found exclusively in rocks just predating the explosion, probably no more than 700 million years old and perhaps younger. Second, the Ediacara animals may represent a failed, independent experiment in multicellular life, not a set of simpler ancestors for later creatures with hard parts. (I shall discuss the nature and status of the Ediacara fauna in chapter V.)

In one sense, the Ediacara fauna poses more problems than it solves for Darwin's resolution of the Cambrian explosion. The most promising version of the "imperfection theory" holds that the Cambrian explosion only marks the appearance of hard parts in the fossil record. Multicellular life may have undergone a long history of gradually ascending complexity leaving no record in the rocks because we have found no "Burgess Shale," or soft-bodied fauna, for the Precambrian. I would not challenge the contribution of this eminently sensible argument to the resolution of the Cambrian enigma, but it cannot provide a full explanation if Ediacara animals are not ancestors for the Cambrian explosion. For the Ediacara creatures *are* soft-bodied, and they are not confined to some odd enclave stuck away in a peculiar Australian environment; they represent a world-wide fauna. So if the true ancestors of Cambrian creatures lacked hard parts, why have we not found them in the abundant deposits that contain the soft-bodied Ediacara fauna?

Puzzles mount upon puzzles the more we consider details of the astounding 100-million-year period between the Ediacara fauna and the consolidation of modern body plans in the Burgess Shale. The beginning of the Cambrian is not marked by the appearance of trilobites and the full range of modern anatomy identified as the Cambrian explosion. The first fauna of hard parts, called the Tommotian after a locality in Russia (but also world-wide in extent), contains some creatures with identifiably modern design, but most of its members are tiny blades, caps, and cups of uncertain affinity—the "small shelly fauna," we paleontologists call it, with honorable frankness and definite embarrassment. Perhaps efficient calcification had not yet evolved, and the Tommotian creatures are ancestors that had not yet developed full skeletons, but only laid down bits of mineralized matter in small and separate places all over their bodies. But perhaps the Tommotian fauna is yet another failed experiment, later supplanted by trilobites and their cohort in the final pulse of the Cambrian explosion.

Thus, instead of Darwin's gradual rise to mounting complexity, the 100 million years from Ediacara to Burgess may have witnessed three radically different faunas—the large pancake-flat soft-bodied Ediacara creatures, the tiny cups and caps of the Tommotian, and finally the modern fauna, culminating in the maximal anatomical range of the Burgess. Nearly 2.5 billion years of prokaryotic cells and nothing else—two-thirds of life's history in stasis at the lowest level of recorded complexity. Another 700 million years of the larger and much more intricate eukaryotic cells, but no aggregation to multicellular animal life. Then, in the 100-million-year wink of a geological eye, three outstandingly different faunas—from Ediacara, to Tommotian, to Burgess. Since then, more than 500 million years of wonderful stories, triumphs and tragedies, but not a single new phylum, or basic anatomical design, added to the Burgess complement.

Step way way back, blur the details, and you may want to read this sequence as a tale of predictable progress: prokaryotes first, then eukaryotes, then multicellular life. But scrutinize the particulars and the comforting story collapses. Why did life remain at stage 1 for two-thirds of its history if complexity offers such benefits? Why did the origin of multicellular life proceed as a short pulse through three radically different faunas, rather than as a slow and continuous rise of complexity? The history of life is endlessly fascinating, endlessly curious, but scarcely the stuff of our usual thoughts and hopes.

LIFE AFTER THE BURGESS: SOFT-BODIED FAUNAS AS WINDOWS INTO THE PAST

An old paleontological in joke proclaims that mammalian evolution is a tale told by teeth mating to produce slightly altered descendant teeth. Since enamel is far more durable than ordinary bone, teeth may prevail when all else has succumbed to the whips and scorns of geological time. The majority of fossil mammals are known only by their teeth.

Darwin wrote that our imperfect fossil record is like a book preserving just a few pages, of these pages few lines, of the lines few words, and of those words few letters. Darwin used this metaphor to describe the chances of preservation for ordinary hard parts, even for maximally durable teeth. What hope can then be offered to flesh and blood amidst the slings and arrows of such outrageous fortune? Soft parts can only be preserved, by a stroke of good luck, in an unusual geological context—insects in amber, sloth dung in desiccated caves. Otherwise, they quickly succumb to the thousand natural shocks that flesh is heir to—death, disaggregation, and decay, to name but three.

And yet, without evidence of soft anatomy, we cannot hope to understand either the construction or the true diversity of ancient animals, for two obvious reasons: First, most animals have no hard parts. In 1978, Schopf analyzed the potential for fossilization of an average modern marine fauna of the intertidal zone. He concluded that only 40 percent of genera could appear in the fossil record. Moreover, potential representation is strongly biased by habitat. About two-thirds of the sessile (immobile) creatures living on the sea floor might be preserved, as contrasted with only a quarter of the burrowing detritus feeders and mobile carnivores. Second, while the hard parts of some creatures—vertebrates and arthropods, for example—are rich in information and permit a good reconstruction of the basic function and anatomy of the entire animal, the simple roofs and coverings of other creatures tell us nearly nothing about their underlying organization. A worm tube or a snail shell implies very little about the organism inside, and in the absence of soft parts, biologists often confuse one for the other. We have not resolved the status of the earth's first multicellular fauna with hard parts, the Tommotian problem (discussed in chapter V), because these tiny caps and covers provide so little information about the creatures underneath.

Paleontologists have therefore sought and treasured soft-bodied faunas since the dawn of the profession. No pearl has greater price in the fossil record. Acknowledging the pioneering work of our German colleagues, we designate these faunas of extraordinary completeness and richness as *Lagerstätten* (literally "lode places," or "mother lodes" in freer translation). *Lagerstätten* are rare, but their contribution to our knowledge of life's history is disproportionate to their frequency by orders of magnitude. When my colleague and former student Jack Sepkoski set out to catalogue the history of all lineages, he found that 20 percent of major groups are known exclusively by their presence in the three greatest Paleozoic *Lagerstätten*—the Burgess Shale, the Devonian Hunsrückschiefer of Germany, and the Carboniferous Mazon Creek near Chicago. (I shall, for the rest of this book, use the standard names of the geological time scale without further explanation. If you spurn, dear reader, my exhortation to memorize this alphabet, please refer to figure 2.1. I also recommend the mnemonics at the beginning of this chapter.)

An enormous literature has been generated on the formation and interpretation of *Lagerstätten* (see Whittington and Conway Morris, 1985). Not all issues have been resolved, and the ins and outs of detail provide endless fascination, but three factors (found in conjunction only infrequently) stand out as preconditions for the preservation of soft-bodied faunas: rapid burial of fossils in undisturbed sediment; deposition in an environment free from the usual agents of immediate destruction—primarily oxygen and other promoters of decay, and the full range of organisms, from bacteria to large scavengers, that quickly reduce most carcasses to oblivion in nearly all earthly environments; and minimal disruption by the later ravages of heat, pressure, fracturing, and erosion.

As one example of the Catch-22 that makes the production of *Lagerstätten* so rare, consider the role of oxygen (see Allison, 1988, for a dissenting view on the importance of anoxic habitats). Environments without oxygen are excellent for the preservation of soft parts: no oxidation, no decay by aerobic bacteria. Such conditions are common on earth, particularly in stagnant basins. But the very conditions that promote preservation also decree that few organisms, if any, make their natural home in such places. The best environments therefore contain nothing to preserve! The "trick" in producing *Lagerstätten*—including the Burgess Shale, as we shall see— lies in a set of peculiar circumstances that can occasionally bring a fauna into such an inhospitable place. *Lagerstätten* are therefore rooted in rarity.

If the Burgess Shale did not exist, we would not be able to invent it, but we would surely pine for its discovery. The Good Lord of Earthly Reality seldom answers our prayers, but he has come through for the Burgess. If Aladdin's djinn had appeared to any paleontologist before the discovery of the Burgess, and stingily offered but one wish, our lucky beneficiary would surely have said without hesitation: "Give me a soft-bodied fauna right after the Cambrian explosion; I want to see what that great episode really produced." The Burgess Shale, our djinn's gift, tells a wonderful story, but not enough for a book by itself. This fauna becomes a key to understanding the history of life by comparison with the strikingly different pattern of disparity in other *Lagerstätten*.

Rarity has but one happy aspect—given enough time, it gets converted to fair frequency. The discovery and study of *Lagerstätten* has accelerated greatly in the past ten years, inspired in part by insights from the Burgess. The total number of *Lagerstätten* is now large enough to provide a good feel for the basic patterns of anatomical disparity through time. If *Lagerstätten* were not reasonably well distributed we would know next to nothing about Precambrian life, for everything from the first prokaryotic cells to the Ediacara fauna is a story of soft-bodied creatures.

As its primary fascination, the Burgess Shale teaches us about an amazing difference between past and present life: with far fewer species, the Burgess Shale—one quarry in British Columbia, no longer than a city block—contains a disparity in anatomical design far exceeding the modern range throughout the world!

Perhaps the Burgess represents a rule about the past, not a special feature of life just after the Cambrian explosion? Perhaps all faunas of such exquisite preservation show a similar breadth of anatomical design? We can only resolve this question by studying temporal patterns of disparity as revealed in other *Lagerstätten.*

The basic answer is unambiguous: the broad anatomical disparity of the Burgess is an exclusive feature of the first explosion of multicellular life. No later *Lagerstätten* approach the Burgess in breadth of designs for life. Rather, proceeding forward from the Burgess, we can trace a rapid stabilization of the decimated survivors. The magnificently preserved, three-dimensional arthropods from the Upper Cambrian of Sweden (Müller, 1983; Müller and Walossek, 1984) may all be members of the crustacean line. (As a result of oddities in preservation, only tiny arthropods, less than two millimeters in length, have been recovered from this fauna, so we can't really compare the disparity in these deposits with the story of larger-bodied Burgess forms.) The Lower Silurian Brandon Bridge fauna from Wisconsin, described by Mikulic, Briggs, and Kluessendorf (1985a and 1985b), contains (like the Burgess) all four major groups of arthropods. It also includes a few oddballs—some unclassifiable arthropods (including one creature with bizarre winglike extensions at its sides) and four worm-like animals, but none so peculiar as the great Burgess enigmas like *Opabinia, Anomalocaris,* or *Wiwaxia.*

The celebrated Devonian Hunsrückschiefer, so beautifully preserved that fine details emerge in X-ray photos of solid rock (Stürmer and Bergström, 1976 and 1978), contains one or two unclassifiable arthropods, including *Mimetaster,* a probable relative of *Marrella,* the most common animal in the Burgess. But life had already stabilized. The prolific Mazon Creek fauna, housed in concretions that legions of collectors have split by the millions over the past several decades, does include a bizarre wormlike animal known as the Tully Monster (officially honored in formal Latin doggerel as *Tullimonstrum*). But the Burgess motor of invention had been shut off by then, and nearly all the beautiful fossils of Mazon Creek fit comfortably into modern phyla.

When we pass through the Permo-Triassic extinction and come to the most famous of all *Lagerstätten*—the Jurassic Solnhofen limestone of Germany—we gain enough evidence to state with confidence that the Burgess game is truly over. No fauna on earth has been better studied. Quarrymen and amateur collectors have been splitting these limestone blocks for more than a century. (These uniform, fine-grained stones are the mainstay of lithography, and have been used, almost exclusively, for all fine prints in this medium ever since the technique was invented at the end of the eighteenth century.) Many of the world's most famous fossils come from these quarries, including all six specimens of *Archaeopteryx,* the first bird, preserved with feathers intact to the last barbule. But the Solnhofen contains nothing, not a single animal, falling outside well-known and well-documented taxonomic groups.

Clearly, the Burgess pattern of stunning disparity in anatomical design is not characteristic of well-preserved fossil faunas in general. Rather, good preservation has permitted us to identify a particular and immensely puzzling aspect of the Cambrian explosion and its immediate aftermath. In a geological moment near the beginning of the Cambrian, nearly all modern phyla made their first appearance, along with an even greater array of anatomical experiments that did not survive very long thereafter. The 500 million subsequent years have produced no new phyla, only twists and turns upon established designs—even if some variations, like human consciousness, manage to impact the world in curious ways. What established the Burgess motor? What turned it off so quickly? What, if anything, favored the small set of surviving designs over other possibilities that flourished in the Burgess Shale? What is this pattern of decimation and stabilization trying to tell us about history and evolution?

THE SETTING OF THE BURGESS SHALE

WHERE

On July 11, 1911, C. D. Walcott's wife, Helena, was killed in a railway accident at Bridgeport, Connecticut. Following a custom of his time and social class, Charles kept his sons close to home, but sent his grieving daughter Helen on a grand tour of Europe, accompanied by a chaperone with the improbable name of Anna Horsey, there to assuage grief and regain composure. Helen, with the enthusiasm of late teen-aged years, did thrill to the monuments of Western history, but she saw nothing to match the beauty of a different West—the setting of the Burgess Shale, where she had accompanied her father both during the discovery of 1909 and the first collecting season of 1910. From Europe, Helen wrote to her brother Stuart in March 1912:

> They have the most fascinating castles and fortresses perched on the very tops. You can just see the enemy creeping up and up—then being surprised by rocks and arrows thrown down on them. We saw, of course, the famous Appian Way and the remains of the old Roman aqueducts—just imagine,

those ruined-looking arches were built nearly 2000 years ago! It makes America seem a little shiny and new, but I'd prefer Burgess Pass to anything I've seen yet.

The legends of fieldwork locate all important sites deep in inaccessible jungles inhabited by fierce beasts and restless natives, and surrounded by miasmas of putrefaction and swarms of tsetse flies. (Alternative models include the hundredth dune after the death of all camels, or the thousandth crevasse following the demise of all sled dogs.) But in fact, many of the finest discoveries, as we shall soon see, are made in museum drawers. Some of the most important natural sites require no more than a pleasant stroll or a leisurely drive; you can almost walk to Mazon Creek from downtown Chicago.

The Burgess Shale occupies one of the most majestic settings that I have ever visited—high in the Canadian Rockies at the eastern border of British Columbia. Walcott's quarry lies at an elevation of almost eight thousand feet on the western slope of the ridge connecting Mount Field and Mount Wapta. Before visiting in August 1987, I had seen many photos of Walcott's quarry; I took several more in the conventional orientation (literally east, looking into the quarry, figure 2.2). But I had not realized the power and beauty of a simple about-face. Turn around to the west, and you confront one of the finest sights on our continent—Emerald Lake below, and the snow-capped President range beyond (figure 2.3), all lit, in late afternoon, by the falling sun. Walcott found some wonderful fossils on the Burgess ridge, but I now have a visceral appreciation of why, well into his seventies, he rode the transcontinental trains year after year, to spend long summers in tents and on horseback. I also understand the appeal of Walcott's principal avocation—landscape photography, including pioneering work in the technology of wide-angle, panoramic shots (figure 2.4).

But the Burgess Shale does not hide in an inaccessible wilderness. It resides in Yoho National Park, near the tourist centers of Banff and Lake Louise. Thanks to the Canadian Pacific Railway, whose hundred-car freights still thunder through the mountains almost continuously, the Burgess Shale lies on the border of civilization. The railroad town of Field (population about 3,000, and probably smaller today than in Walcott's time, especially since the Railway hotel burned down) lies just a few miles from the site, and you can still board the great transcontinental train from its tiny station.

Today you can drive to the Takakkaw Falls campground, near the Whiskey Jack Hostel (named after a bird, not an inebriated hero of the old West), and then climb the three thousand feet up to Burgess Ridge by way

A   B

C

2.2.   Three views of the Burgess Shale quarries taken during my visit in August 1987. (A) The northern end of Walcott's quarry, with Mount Wapta in the background. Note the quarry wall with cores drilled for the insertion of dynamite charges, and the debris from blasting on the quarry floor. (B) A similar view of the quarry opened by Percy Raymond in 1930, with yours truly and three avid geologists. This much smaller quarry lies above Walcott's original site. (C) My son Ethan sitting on the floor of Walcott's quarry as seen at the southern end.

2.3.   The view from Walcott's quarry. A geologist searches for fossils on the talus slope in the foreground. Emerald Lake lies beyond.

2.4.   This reduced version of one of Walcott's famous panoramic photographs gives a good impression of the technique, but lacks the grandeur of the original, which is several feet long. Walcott took this photograph in 1913. The right-hand side shows the Burgess quarry, with Mount Wapta to the left. Note some collectors and collecting tools within the quarry.

of a four-mile trail around the northwest flank of Mount Wapta. The climb has some steep moments, but it qualifies as little more than a pleasant stroll, even for yours truly, overweight, out of shape, and used to life at sea level. A more serious field effort can now employ helicopters to fly supplies in and out (as did the Geological Survey of Canada expeditions of the 1960s and the Royal Ontario Museum parties of the 1970s and 1980s). Walcott had to rely upon pack horses, but no one could brand the effort as overly strenuous or logistically challenging, as field work goes. Walcott himself (1912) provided a lovely description of his methods during the first field season of 1910—a verbal snapshot that folds an older technology and social structure into its narrative, with active sons scouring the hillside and a dutiful wife trimming the specimens back at camp:

> Accompanied by my two sons, Sidney and Stuart . . . we finally located the fossil-bearing band. After that, for days we quarried the shale, slid it down the mountain side in blocks to a trail, and transported it to camp on pack horses, where, assisted by Mrs. Walcott, the shale was split, trimmed and packed, and then taken down to the railway station at Field, 3,000 feet below.

A year before he discovered the Burgess Shale, Walcott (1908) described an equally charming, rustic technology for collecting from the famous *Ogygopsis* trilobite bed of Mount Stephen, a locality similar in age to the Burgess, and just around the next bend:

> The best way to make a collection from the "fossil bed" is to ride up the trail on a pony to about 2,000 feet above the railroad, collect specimens, securely wrap them in paper, place them in a bag, tie the bag to the saddle, and lead the pony down the mountain. A fine lot can be secured in a long day's trip, 6:00 AM to 6:00 PM.

The romance of the Burgess has had at least one permanent effect upon all future study of its fossils—the setting of their peculiar names. The formal Greek and Latin names of organisms can sometimes rise to the notable or the mellifluous, as in my favorite moniker, for a fossil snail—*Pharkidonotus percarinatus* (say it a few times for style). But most designations are dry and literal: the common rat is, for overkill, *Rattus rattus rattus;* the two-horned rhino is *Diceros;* the periwinkle, an inhabitant of nearshore, or littoral, waters, is *Littorina littorea.*

Burgess names, by contrast, are a strange-sounding lot. Decidedly not Latin in their roots, they are sometimes melodious, as in *Opabinia,* but other times nearly unpronounceable for their run of vowels, as in *Aysheaia, Odaraia,* and *Naraoia,* or their unusual consonants, as in *Wiwaxia, Takak-*

kawia, and *Amiskwia.* Walcott, who loved the Canadian Rockies and spent a quarter century of summers in its field camps, labeled his fossils with the names of local peaks and lakes,* themselves derived from Indian words for weather and topography. *Odaray* means "cone-shaped"; *opabin* is "rocky"; *wiwaxy,* "windy."

WHY: THE MEANS OF PRESERVATION

Walcott found almost all his good specimens in a lens of shale, only seven or eight feet thick, that he called the "phyllopod bed." ("Phyllopod," from the Latin for "leaf-footed," is an old name for a group of marine crustaceans bearing leaflike rows of gills on one branch of their legs. Walcott chose this name to honor *Marrella,* the most common of Burgess organisms. Citing the numerous rows of delicate gills, Walcott dubbed *Marrella* the "lace crab" in his original field notes. According to later studies, *Marrella* is neither crab nor phyllopod, but one of the taxonomically unique arthropods of the Burgess Shale.)

At this level, fossils are found along less than two hundred feet of outcrop on the modern quarry face. Since Walcott's time, additional soft-bodied fossils have been collected at other stratigraphic levels and localities in the area. But nothing even approaching the diversity of the phyllopod bed occurs anywhere else, and Walcott's original layer has yielded the great majority of Burgess species. Little taller than a man, and not so long as a city block! When I say that one quarry in British Columbia houses more anatomical disparity than all the world's seas today, I am speaking of a *small* quarry. How could such richness accumulate in such a tiny space?

Recent work has clarified the geology of this complex area, and provided a plausible scenario for deposition of the Burgess fauna (Aitken and McIlreath, 1984; and the more general discussion in Whittington, 1985b). The animals of the Burgess Shale probably lived on mud banks built up along the base of a massive, nearly vertical wall, called the Cathedral Escarpment—a reef constructed primarily by calcareous algae (reef-building corals had not yet evolved). Such habitats in moderately shallow water, adequately lit and well aerated, generally house typical marine faunas of high diversity. The Burgess Shale holds an ordinary fauna from habitats well represented in the fossil record. We cannot attribute its extraordinary disparity of anatomical designs to any ecological oddity.

---

*Burgess himself was a nineteenth-century governor general of Canada; Walcott named the formation not for him but for Burgess Pass, which provided access to the quarry from the town of Field.

Catch-22 now intrudes. The very typicality of the Burgess environment should have precluded any preservation of a soft-bodied fauna. Good lighting and aeration may encourage high diversity, but should also guarantee rapid scavenging and decay. To be preserved as soft-bodied fossils, these animals had to be moved elsewhere. Perhaps the mud banks heaped against the walls of the escarpment became thick and unstable. Small earth movements might have set off "turbidity currents" propelling clouds of mud (containing the Burgess organisms) down slope into lower adjacent basins that were stagnant and devoid of oxygen. If the mudslides containing Burgess organisms came to rest in these anoxic basins, then all the factors for overcoming Catch-22 fall into place—movement of a fauna from an environment where soft anatomy could not be preserved to a region where rapid burial in oxygen-free surroundings could occur. (See Ludvigsen, 1986, for an alternate view that preserves the central idea of burial in a relatively deep-water anoxic basin, but replaces a slide of sediments down an escarpment with deposition at the base of a gently sloping ramp.)

The pinpoint distribution of the Burgess fossils supports the idea that they owe their preservation to a local mudslide. Other features of the fossils lead to the same conclusion: very few specimens show signs of decay, implying rapid burial; no tracks, trails, or other marks of organic activity have been found in the Burgess beds, thus indicating that the animals died and were overwhelmed by mud as they reached their final resting place. Since nature usually sneezes on our hopes, let us give thanks for this rare concatenation of circumstances—one that has enabled us to wrest a great secret from a generally uncooperative fossil record.

WHO, WHEN: THE HISTORY OF DISCOVERY

Since this book is a chronicle of a great investigation that reversed Walcott's conventional interpretation of the Burgess fossils, I find it both fitting in the abstract, and beautifully symmetrical in the cause of narrative, that the traditional tale about his discovery is also a venerable legend badly in need of revision.

We are storytelling animals, and cannot bear to acknowledge the ordinariness of our daily lives (and even of most events that, in retrospect, seem crucial to our fortunes or our history). We therefore retell actual events as stories with moral messages, embodying a few limited themes that narrators through the ages have cultivated for their power to interest and to instruct.

The canonical story for the Burgess Shale has particular appeal because it moves gracefully from tension to resolution, and enfolds within its basically simple structure two of the greatest themes in conventional narration—serendipity and industry leading to its just reward.* Every paleontologist knows the tale as a staple of campfires and as an anecdote for introductory courses. The traditional version is best conveyed by an obituary for Walcott written by his old friend and former research assistant Charles Schuchert, professor of paleontology at Yale:

> One of the most striking of Walcott's faunal discoveries came at the end of the field season of 1909, when Mrs. Walcott's horse slid on going down the trail and turned up a slab that at once attracted her husband's attention. Here was a great treasure—wholly strange Crustacea of Middle Cambrian time—but where in the mountain was the mother rock from which the slab had come? Snow was even then falling, and the solving of the riddle had to be left to another season, but next year the Walcotts were back again on Mount Wapta, and eventually the slab was traced to a layer of shale—later called the Burgess shale—3000 feet above the town of Field (1928, pp. 283–84).

Consider the primal character of this tale—the lucky break provided by the slipping horse (figure 2.5), the greatest discovery at the very last minute of a field season (with falling snow and darkness heightening the drama of finality), the anxious wait through a winter of discontent, the triumphant return and careful, methodical tracing of errant block to mother lode. Schuchert doesn't mention a time for this last act of patient discovery, but most versions claim that Walcott spent a week or more trying to locate the source of the Burgess Shale. His son Sidney, reminiscing sixty years later, wrote (1971, p. 28): "We worked our way up, trying to find the bed of rock from which our original find had been dislodged. A week later and some 750 feet higher we decided that we had found the site."

A lovely story, but none of it is true. Walcott, a great conservative administrator (see chapter IV), left a precious gift to historians in his meticulous habits of assiduous record keeping. He never missed a day in his diary, and we can reconstruct the events of 1909 with fair precision. Walcott found the first soft-bodied fossils on Burgess Ridge on either August 30 or 31. His entry for August 30 reads:

> Out collecting on the Stephen formation [the larger unit that includes what Walcott later called the Burgess Shale] all day. Found many interesting

*Much material in this section comes from my previous essay on Walcott's discovery (Gould, 1988).

2.5. Walcott in his seventies, during one of his last Western field seasons. He stands with his horse, reminding us of the legend of the discovery of the Burgess Shale.

> fossils on the west slope of the ridge between Mounts Field and Wapta [locality of the Burgess Shale]. Helena, Helen, Arthur and Stuart [his wife, daughter, assistant, and son] came up with remainder of outfit at 4 P.M.

The next day, they had obviously discovered a rich assemblage of soft-bodied fossils. Walcott's quick sketches (figure 2.6) are so clear that I can identify the three genera depicted: *Marrella* (upper left), one of the unclassifiable arthropods; *Waptia* (upper right); and the peculiar trilobite *Naraoia* (lower left). Walcott wrote: "Out with Helena and Stuart collecting fossils from the Stephen formation. We found a remarkable group of phyllopod crustaceans. Took a large number of fine specimens to camp."

What about the horse slipping and the snow falling? If this incident occurred at all, it must have been on August 30, when his family came up the slope to meet him in the late afternoon. They might have turned up the slab as they descended for the night, returning the next morning to find the specimens that Walcott sketched on August 31. This reconstruction gains some support from a letter that Walcott wrote to Marr (for whom he later named the "lace crab" *Marrella*) in October 1909:

> When we were collecting from the Middle Cambrian, a stray slab brought down by a snow slide showed a fine phyllopod crustacean on a broken edge. Mrs. W. and I worked on that slab from 8 in the morning until 6 in the

evening and took back with us the finest collection of phyllopod crustaceans that I have ever seen.

Transformation can be subtle. A previous snowslide becomes a present snowstorm, and the night before a happy day in the field becomes a forced and hurried end to an entire season. But, far more importantly, Walcott's field season did not finish with the discoveries of August 30 and 31. The party remained on Burgess ridge until September 7. Walcott was thrilled by his discovery, and he collected with avidity every single day thereafter. Moreover, although Walcott assiduously reported the weather in every entry, the diary breathes not a single word about snow. His happy week brought nothing but praise for Mother Nature. On September 1, he wrote: "Beautiful warm days."

Finally, I strongly suspect that Walcott located the source of his stray block during that last week of 1909—at least the basic area of outcrop, if not the phyllopod bed itself. On September 1, the day after he sketched

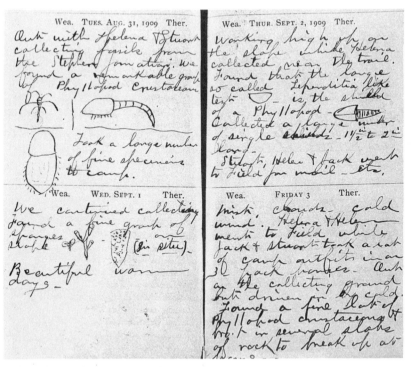

2.6. The smoking gun that disproves the canonical story for the discovery of the Burgess Shale. Walcott sketched three Burgess genera on August 31 and then continued to collect with great success for another week.

the three arthropods, Walcott wrote: "We continued collecting. Found a fine group of sponges on slope (in situ) [that is, undisturbed and in their original position]." Sponges, containing some hard parts, extend beyond the richest layers of soft-bodied preservation at this site, but the best specimens come from the phyllopod bed. On each subsequent day, Walcott found abundant soft-bodied specimens, and his descriptions do not read like the work of a man encountering a lucky stray block here and there. On September 2, he discovered that the supposed shell of an ostracode had really housed the body of a phyllopod: "Working high up on the slope while Helena collected near the trail. Found that the large so called Leperditia like test is the shield of a phyllopod." The Burgess quarry is "high up on the slope," while stray blocks would slide down to the trail.

On September 3, Walcott was even more successful: "Found a fine lot of Phyllopod crustaceans and brought in several slabs of rock to break up at camp." In any event, he continued to collect, and put in a full day for his last hurrah on September 7: "With Stuart and Mr. Rutter went up on fossil beds. Out from 7 A.M. to 6:30 P.M. Our last day in camp for 1909."

If I am right about his discovery of the main bed in 1909, then the second part of the canonical tale—the week-long patient tracing of errant block to source in 1910—must be equally false. Walcott's diary for 1910 supports my interpretation. On July 10, champing at the bit, he hiked up to the Burgess Pass campground, but found the area too deep in snow for any excavations. Finally, on July 29, Walcott reported that his party set up "at Burgess Pass campground of 1909." On July 30, they climbed neighboring Mount Field and collected fossils. Walcott indicates that they made their first attempt to map the Burgess beds on August 1: "All out collecting the Burgess formation until 4 P.M. when a cold wind and rain drove us into camp. Measured section of the Burgess formation—420 feet thick. Sidney with me. Stuart with his mother and Helen puttering about camp." "Measuring a section" is geological jargon for tracing the vertical sequence of strata and noting the rock types and fossils. If you wished to find the source of an errant block that had broken off and tumbled down, you would measure the section above, trying to match your block to its most likely layer.

I think that Charles and Sidney Walcott located the phyllopod bed on this very first day, because Walcott wrote for his next entry, of August 2: "Out collecting with Helena, Stuart and Sidney. We found a fine lot of 'lace crabs' and various odds and ends of things." "Lace crab" was Walcott's field term for *Marrella*, chief denizen of the phyllopod bed. If we wish to give the canonical tale all benefit of doubt, and argue that these "lace crabs" of August 2 came from dislodged blocks, we still cannot grant a week of strenuous effort for locating the mother lode, for Walcott wrote just two days later, on August 4: "Helena worked out a lot of Phyllopod crustaceans from 'Lace Crab' layer."

The canonical tale is more romantic and inspiring, but the plain factuality of the diary makes more sense. The trail lies just a few hundred feet below the main Burgess beds. The slope is simple and steep, with strata well exposed. Tracing an errant block to its source should not have been a major problem, for Walcott was more than a good geologist—he was a great geologist. He should have located the main beds right away, in 1909, in the week after he first discovered the soft-bodied fossils. He did not have an opportunity to quarry in 1909—the only constraint imposed by limits of time—but he found many fine fossils, and probably the main beds themselves. In 1910, he knew just where to go, and he set up shop in the right place as soon as the snow melted.

Walcott established his quarry in the phyllopod bed of the Burgess Shale and worked with hammers, chisels, long iron bars, and small explosive charges for a month or more in each year from 1910 through 1913. In 1917, at age sixty-seven, he returned for a final fifty days of collecting. In all, he brought some eighty thousand specimens back to Washington, D.C., where they still reside, the jewel of our nation's largest collection of fossils, in the National Museum of Natural History at the Smithsonian Institution.

Walcott collected with zeal and thoroughness. He loved the West and viewed his annual trips as a necessary escape for sanity from the pressures of administrative life in Washington. But back at the helm of his sprawling administrative empire, he never found even the entering wedge of ample time to examine, ponder, ruminate, observe again, obsess, reconsider, and eventually publish—the essential (and incompressible) ingredients of a proper study of these complex and precious fossils. (The significance of this failure will emerge as an important theme in chapter IV.)

Walcott did publish several papers with descriptions of Burgess fossils that he labeled "preliminary"—in large part to exercise his traditional right to bestow formal taxonomic names upon his discoveries. Four such papers appeared in 1911 and 1912 (see Bibliography)—the first on arthropods that he considered (incorrectly) as related to horseshoe crabs, the second on echinoderms and jellyfish (probably all attributed to the wrong phyla), the third on worms, and the fourth and longest on arthropods. He never again published a major work on Burgess metazoans. (A 1918 article on trilobite appendages relies largely on Burgess materials. His 1919 work on Burgess algae, and his 1920 monograph on Burgess sponges, treat different taxonomic groups and do not address the central issue of disparity in the anatomical design of coelomate animals. Sponges are not related to other animals and presumably arose independently, from unicellular ancestors. The 1931 compendium of additional descriptions, published under Walcott's name, was compiled after his death by his associate Charles E. Resser from notes that Walcott had never found time to polish and publish.)

In 1930, Percy Raymond, professor of paleontology at Harvard, took three students to the Burgess site and reopened Walcott's old quarry. He also developed a much smaller quarry at a new site just sixty-five feet above Walcott's original. He found only a few new species, but made a fine, if modest, collection.

These specimens—primarily Walcott's, with a small infusion from Raymond—formed the sole basis for all study of the Burgess Shale before Whittington and colleagues began their revision in the late 1960s. Given the supreme importance of these fossils, the amount of work done must be judged as relatively modest, and none of the papers even hint at an interpretation basically different from Walcott's view that the Burgess organisms could all be accommodated within the taxonomic boundaries of successful modern phyla.

I well remember my first encounter with the Burgess Shale, when I was a graduate student at Columbia in the mid-1960s. I realized how superficially Walcott had described these precious fossils, and I knew that most had never been restudied. I dreamed, before I understood my utter lack of administrative talent or desire, about convening an international committee of leading taxonomic experts on all phyla represented in the Burgess. I would then farm out *Amiskwia* to the world's expert on chaetognaths, *Aysheaia* to the dean of onychophoran specialists, *Eldonia* to Mr. Sea Cucumber. None of these taxonomic attributions has stood the test of subsequent revision, but my dream certainly reflected the traditional view propagated by Walcott and never challenged—that all Burgess oddities could be accommodated in modern groups.

Since one cannot set out deliberately to find the unexpected, the work that prompted our radical revision had modest roots. The Geological Survey of Canada, in the course of a major mapping program, was working in the southern Rocky Mountains of Alberta and British Columbia in the mid-1960s. This general effort almost inevitably suggested a reexamination of the Burgess Shale, the most famous site in the region. But no one anticipated any major novelty. Harry Whittington got the nod as paleontologist-in-chief because he was one of the world's leading experts on fossil arthropods—and everyone thought that most of the Burgess oddities were members of this great phylum.

My friend Digby McLaren, then head of the Geological Survey and chief instigator of the Burgess restudy, told me in February 1988 that he had pushed the project primarily for (quite proper) chauvinistic reasons, not from any clear insight about potential intellectual reward. Walcott, an American, had found the most famous Canadian fossils and carted the entire booty back to Washington. Many Canadian museums didn't own a single specimen of their geological birthright. McLaren, declaring this situation a "national shame," set forth (in his only partially facetious words) "to repatriate the Burgess Shale."

For six weeks in the summers of 1966 and 1967, a party of ten to fifteen scientists, led by Harry Whittington and the geologist J. D. Aitken, worked in Walcott's and Raymond's quarries. They extended Walcott's quarry some fifteen meters northward and split about seven hundred cubic meters of rock in Walcott's and seventeen in Raymond's quarry. Besides substituting helicopters for horses and using smaller explosive charges (to avoid jumbling stratigraphic information by throwing fossiliferous blocks too far from their source for proper identification), these modern expeditions worked pretty much as Walcott had. The greatest invention since Walcott, as Whittington notes (1985b, p. 20), is the felt-tipped pen—a godsend for labeling each rock immediately upon collection.

In 1975, Des Collins of the Royal Ontario Museum mounted an expedition to collect fossils from the debris slopes in and around both quarries. He was not permitted to blast or excavate in the quarries themselves, but his party found much valuable material. (The Burgess Shale is so rich that some remarkable novelties could still be found in Walcott's spoil heaps.) In 1981 and 1982, Collins explored the surrounding areas, and found more than a dozen new sites with fossils of soft-bodied organisms in rocks of roughly equivalent age. None approach the Burgess in richness, but Collins has made some remarkable discoveries, including *Sanctacaris,* the first chelicerate arthropod. If Walcott's phyllopod bed arose when a turbidity current triggered a mudslide, then many other similar slides must have occurred at about the same time, and other *Lagerstätten* should abound. As I write this book in the summer of 1988, Des Collins is out searching for more sites in the Canadian Rockies.

Paleontology is a small and somewhat incestuous profession. The Burgess Shale has always stood over my world like a colossus. Bill Schevill, the last survivor of Raymond's 1930 expedition and later a great expert on whales, stops by my office for a chat now and then. G. Evelyn Hutchinson, who described the strange *Aysheaia* and the equally enigmatic *Opabinia* in 1931 (getting one basically right and the other equally wrong), and who later became the world's greatest ecologist and my own intellectual guru, has regaled me with stories about his foray, as a young zoologist, into the peculiar world of fossils. Percy Raymond's collection sits in two large cabinets right outside my office. I was first appointed to Harvard as a very junior replacement for Harry Whittington, who had just taken the chair in geology at Cambridge (where he studied the Burgess for the next twenty years on a transoceanic shuttle). I am no expert on older rocks or the anatomy of arthropods, but I cannot escape the Burgess Shale. It is an icon and symbol of my profession, and I write this book to pay my respects, and to discharge an intellectual debt for the thrill that such creatures can inspire in a profession that might reinterpret Quasimodo's lament as an optimistic plea for fellowship: Oh why was I not made of stone like these!

CHAPTER III

Reconstruction of the Burgess Shale: Toward a New View of Life

A Quiet Revolution

Some transformations are overt and heroic; others are quiet and uneventful in their unfolding, but no less significant in their outcome. Karl Marx, in a famous statement, compared his social revolution to an old mole burrowing busily beneath the ground, invisible for long periods, but undermining traditional order so thoroughly that a later emergence into light precipitates a sudden overturn. But intellectual transformations often remain under the surface. They ooze and diffuse into scientific consciousness, and people may slowly move from one pole to another, having never heard the call to arms. The new interpretation of the Burgess Shale ranks among the most invisible of transformations for two basic reasons, but its power to alter our view of life cannot be matched by any other paleontological discovery.

First, the Burgess revision is an intensely intellectual drama—not a swashbuckling tale of discovery in the field, or of personal struggle to the rhetorical death waged by warring professionals battling for the Nobel gold. The new view trickled forth, tentatively at first but with more confidence later on, in a series of long and highly technical taxonomic and anatomical monographs, published mostly in the *Philosophical Transactions of the Royal Society, London,* the oldest scientific journal in English (dating back to the 1660s), but scarcely an item on the shelf of your corner drugstore, or even your local library, and not the sort of publication scrutinized by the journalists responsible for selecting the tiny part of scientific activity destined for public notice.

Second, all the standard images of scientific discovery were violated by the revision of the Burgess Shale. All the romantic legends about field work, all the technocratic myths about machine-based novelty in procedure, were fractured or simply bypassed.

The myth of field work, for example, proclaims that great alterations in ideas arise from new, pristine discoveries. At the end of the trail, after weeks of blood, sweat, toil, and tears, the intrepid scientist splits a rock from the most inaccessible place on the map, and cries Eureka! as he spies the fossil that will shake the world. Since the Burgess revision was preceded by two full seasons of field work, in 1966 and 1967, most people would assume that discoveries of this expedition prompted the reinterpretation. Well, Whittington and company did find some wonderful specimens, and a few new species, but old Walcott, a maniacal collector, had been there first, and had worked for five full seasons. He therefore got most of the goodies. The expeditions of 1966 and 1967 did spur Whittington into action, but the greatest discoveries were made in museum drawers in Washington—by restudying Walcott's well-trimmed specimens. The greatest bit of "field work," as we shall see, occurred in Washington during the spring of 1973, when Whittington's brilliant and eclectic student Simon Conway Morris made a systematic search through *all* the drawers of Walcott's specimens, consciously looking for oddities because he had grasped the germ of the key insight about Burgess disparity.

The myth of the laboratory invokes the same misconception, transferred indoors—that new ideas must arise from pristine discoveries. According to this "frontier mentality," one can advance only by "seeing the unseen"— by developing some new method to discern what, in principle, could not be perceived before. Progress therefore requires that the boundaries of complex and expensive machinery be extended. Novelty becomes linked inextricably with miles of glassware, banks of computers, cascading numbers, spinning centrifuges, and big, expensive research teams. We may have come a long way from those wonderful Art Deco sets of the old horror films, where Baron Frankenstein harnessed the power of lightning to quicken his monsters, but the flashing lights, tiers of buttons, and whirling dials of that enterprise neatly captured a myth that has only grown since then.

The Burgess revision did require a definite set of highly specialized methods, but the tools of this particular technology do not extend beyond ordinary light microscopes, cameras, and dental drills. Walcott missed some crucial observations because he didn't use these methods—but he could have employed all Whittington's techniques, had he ever found time to ponder, and to recognize their importance. Everything that Whittington did to see farther and better could have been done in Walcott's day.

The actual story of the Burgess may reflect science as practiced, but this basic truthfulness doesn't make my job any easier. Mythology does have its use as a powerful aid to narrative. Yet, after considering many possible modes of composition, I finally decided that I could present this information in only one way. The revision of the Burgess Shale is a drama, however devoid of external pomp and show—and dramas are stories best told in chronological order. This chapter, the centerpiece of my book, shall therefore proceed as a narrative in proper temporal sequence (preceded by an introduction on methods of study and followed by discussion of the wider implications).

But how to establish chronology? The obvious method of simply asking the major players for their memories cannot suffice. Oh, I did my duty in this regard. I visited them all, pad and pencil in hand. The exercise made me feel rather foolish, for I know these men well, and we have been discussing the Burgess Shale over beer and coffee for nearly two decades.

Besides, the worst possible source for what Harry Whittington thought in 1971, when he published his first monograph on *Marrella*, is Harry Whittington in 1988. How can one possibly peel away an entire edifice of later thought to recover an embryonic state of mind unaffected by the daily intellectual struggles of nearly twenty subsequent years? The timing of events becomes jumbled in retrospect, for we arrange our thoughts in a logical or psychological order that makes sense to us, not in chronological sequence.*

I call this the "my, how you've grown" phenomenon. No comment from relatives is more universally detested by children. But the relatives are correct; they haven't visited for a long time and do accurately remember the last meeting long ago, while a child sees his past dimly through all the intervening events. Freud once remarked that the human mind is like a psychic Rome in violation of the physical law that two objects can't occupy the same space at the same time. No buildings are demolished, and structures from the time of Romulus and Remus join the restored Sistine Chapel in a confusing jumble that also heaps the local trattoria upon the

---

*I know this so well from personal experience. People ask me all the time what I was thinking when Niles Eldredge and I first developed the theory of punctuated equilibrium in the early 1970s. I tell them to read the original paper, for I don't remember (or at least cannot find those memories amidst the jumble of my subsequent life).

Roman bath. The recovery of chronological order requires contemporary documents.

I have therefore worked primarily from the published record. My procedure was simplicity itself. I read technical monographs in strictly chronological order, focusing almost entirely on the primary works of anatomical description, not the fewer articles of secondary interpretation. I may be a lousy reporter, but at least I can proceed as no journalist or "science writer" ever would, or could. The men who revised the Burgess Shale are my colleagues, not my subjects. Their writings are my literature, not the distant documents of another world. I read more than a thousand pages of anatomical description, loving every word—well, most of them at least—and knowing by personal practice exactly how the work had been done. I started with Whittington's first monograph on *Marrella* (1971), and only stopped when I ran out at *Anomalocaris* (Whittington and Briggs, 1985), *Wiwaxia* (Conway Morris, 1985), and *Sanctacaris* (Briggs and Collins, 1988). I don't know that I have ever had more fun, or experienced more appreciation for exquisite work beautifully done, than during the two months that I devoted to this exercise.

Does such a procedure distort or limit the description of science? Of course it does. Every scientist knows that most activities, particularly the mistakes and false starts, don't enter the published record, and that conventions of scientific prose would impart false views of science as actually done, if we were foolish enough to read technical papers as chronicles of practice. Bearing this self-evident truth in mind, I shall call upon a variety of sources as I proceed. But I prefer to focus on the monographic record for a particular, and largely personal, reason.

The psychology of discovery is endlessly fascinating, and I shall not ignore that subject. But the logic of argument, as embodied in published work, has its own legitimate, internal appeal. You can pull an argument apart into its social, psychological, and empirical sources—but you can also cherish its integrity as a coherent work of art. I have great respect for the first strategy, the mainstay of scholarship, but I love to practice the second as well (as I did in my book *Time's Arrow, Time's Cycle*, an analysis of the central logic in three texts crucial for geology's discovery of time). Chronological change in a succession of arguments, each coherent at its own moment, forms a primary record of intellectual development.

The revision of the Burgess Shale involves hundreds of people, from the helicopter pilots who flew supplies in and out of Burgess base camp, to the draftsmen and artists who prepared drawings for publication, to an international group of paleontologists who offered support, advice, and criticism. But the research program of monographic revision has centered on one coherent team. Three people have played the focal role in these efforts: the originator of the project and chief force throughout, Harry Whittington, professor of geology at Cambridge University (that is, in British terminology, senior figure and department head), and two men who began as graduate students under him in the early 1970s and have since built brilliant careers on their researches in the Burgess Shale—Simon Conway Morris (now also at Cambridge) and Derek Briggs (now at Bristol University). Whittington also collaborated with two junior colleagues, especially before his graduate students arrived—Chris Hughes and David Bruton.

The seeds of conventional drama lie with these people, particularly in the interaction between Whittington and Conway Morris, but I have no such story to tell. Whittington is meticulous and conservative, a man who follows the paleontological straight and narrow, eschewing speculation and sticking to the rocks—exactly the opposite of anyone's image for an agent of intellectual transformation. Conway Morris, before the inevitable mellowing of ontogeny, was a fiery Young Turk, a social radical of the 1970s. He is, by temperament, a man of ideas, but happily possessed of the patience and *Sitzfleisch* needed to stare at blobs on rocks for hours on end. In legend, the Burgess reinterpretation would have emerged as a tense synergism between these men—Harry instructing, pleading caution, forcing attention to the rocks; Simon exhorting, pushing for intellectual freedom, nudging his reluctant old mentor toward a new light. One can imagine the discussions, the escalating arguments, the threats, the near fracturings, the break, the return of the prodigal son, the reconciliation.

I don't think that any of this occurred, at least not overtly. And if you know the British university system, you will immediately understand why. British doctoral students study in nearly complete independence. They take no courses, but only work on their dissertation. They agree on a topic with their mentor, and then start their research. If they are lucky, they may meet with their adviser once every month or so; once a year would be more likely. Harry Whittington, a quiet, conservative, and inordinately busy man, was not about to challenge this peculiar tradition. Simon has told me that "Harry didn't like being disturbed," for he "grudged every moment that he couldn't get on with his research." But he was, Simon insists, "a splendid adviser; for he left us alone and he got us support."

I have questioned Harry, Simon, and Derek many times, trying to probe through my initial disbelief. They all insist that they never viewed themselves as a team with a coherent purpose or a general attitude. They were not striving actively to develop a central interpretation together. They never met regularly; in fact, they insist that they never met as a group at all. They didn't even encounter each other on the one certain gathering ground of any British academic department—the almost unmissable daily ritual of morning coffee—for Simon, the social radical, had formed a rump group in his office, and never came, while Harry, who could always see essence beneath externality (the key to deciphering the Burgess animals, after all), never insisted on conformity of any kind. Oh, they all engaged in complex cross-fertilization—but as much, I suspect, by reading each other's papers as by any programmatic or regular discussion. The most I could wrest from any of the trio was an acknowledgement by Derek Briggs that they developed "some corporate perception, even if not by daily interaction."

The drama I have to tell is intense and intellectual. It transcends these ephemeral themes of personality and the stock stage. The victory at stake is bigger and far more abstract than any material reward—a new interpretation of life's history. This goal, once achieved, brings no particular earthly benefit. Paleontology has no Nobel prizes—though I would unhesitatingly award the first to Whittington, Briggs, and Conway Morris as a trio. And, as the old clichés go, you can't fry an egg with your new view of life, or get on the subway, unless you also have a token. (I don't think it even gets you any frequent-flyer miles, though almost everything else does.) You do get the gratitude of your fellow paleontologists, and it doesn't harm your job prospects. But the main reward must be satisfaction—the privilege of working on something exciting, the internal peace of accomplishment, the rare pleasure of knowing that your life made a difference. What more can a person want than to hear, from whatever source he honors as absolute and permanent, the ultimate affirmation that life has been useful: "Well done, thou good and faithful servant"?

A METHODOLOGY OF RESEARCH

A common misconception holds that soft-bodied fossils are usually preserved as flat films of carbon on the surface of rocks. The Burgess organisms are, of course, strongly compressed—we cannot expect the preservation of much three-dimensional structure as the weight of water and sediment piles above an entombed body devoid of hard parts. But the Burgess fossils are not always completely flattened—and this discovery provided Whittington with the basis for a method that could reveal their structure. (Burgess soft parts, by the way, are not preserved as carbon. By a chemical process not yet understood, the original carbon was replaced by silicates of alumina and calcium, forming a dark reflective layer. This replacement did not compromise the exquisite preservation of anatomical detail.)

Walcott never recognized, or appreciated only dimly, that some three-dimensional architecture had been retained. He treated the Burgess fossils as flat sheets, and therefore worked by searching through his specimens for the ones preserved in the most revealing (or least confusing) orientation—usually, for bilaterally symmetrical animals, splayed out straight and flat (as in figure 3.1, a typical Walcott illustration). He ignored specimens in an oblique or frontal orientation, because he thought that the different parts and surfaces so encountered would be squashed together into a single uninterpretable film on the bedding plane; a top view, by contrast, would offer maximal resolution of separate features.

Walcott illustrated his specimens by photographs, often egregiously retouched. Whittington's group has also used photography extensively, but mostly for publication, rather than as a primary research tool. The Burgess specimens do not photograph well (figure 3.2 is a magnificent exception), and little can be gained by working from prints, however enlarged or filtered, rather than from actual specimens. The aluminosilicate surfaces reflect light in various ways at different angles of illumination—and some resolution has been gained by comparing the dull images obtained at high angles of illumination with the bright reflections, obtained at low angles.

Whittington therefore used the oldest method of all as his primary mode of illustration—patient and detailed drawing of specimens. The basic item of machinery, the camera lucida, is no different now from the model that Walcott used, and not much improved from its original invention by the mineralogist W. H. Wollaston in 1807. A camera lucida is, basically, a set of mirrors that can focus the image of an object onto a flat surface. You can attach a camera lucida to a microscope and cast the image under the lens onto a piece of paper. By simultaneously viewing the specimen and its reflection on paper, you can draw the animal without moving your head from the eyepiece. Whittington and his team adopted the procedure of drawing every specimen, at very large scale, for any species under investigation. You can study a set of drawings together, but you cannot easily make simultaneous observations on numerous tiny specimens, all needing magnification.

Whittington applied his camera lucida and skill in drafting to a set of methods all linked to his central recognition that the Burgess fossils retained some three-dimensional structure, and were not just flattened sheets on bedding planes. I shall illustrate the power of these simple procedures by showing their usefulness in the study of the largest Burgess arthropod, the species that Walcott named *Sidneyia inexpectans* to honor his son,

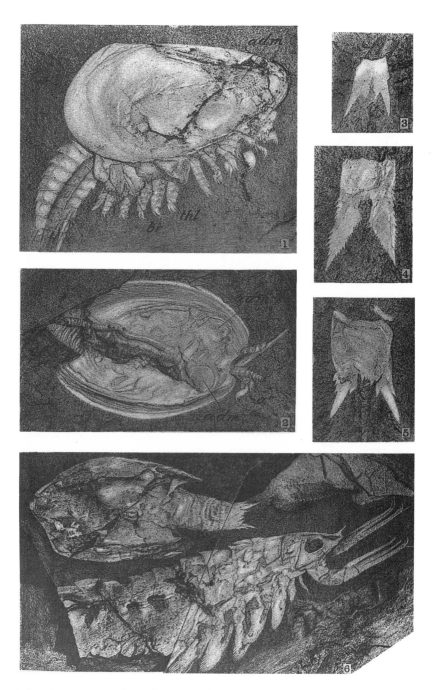

3.1. An attractive plate of Burgess photographs from Walcott's 1912 monograph on arthropods. The photographs are extensively retouched. *Canadaspis* is at top left; *Leanchoilia* at bottom.

3.2. The best unretouched photo ever taken of a Burgess Shale organism. Des Collins took this photograph of a *Naraoia,* preserved in side view. This specimen does not come from Walcott's quarry, but from one of the dozen additional localities for soft-bodied fossils recently found by Collins in the same area. Specimens from Walcott's quarry do not photograph this well.

who had found the first specimen. (I choose *Sidneyia* because David Bruton's 1981 monograph on this genus is, in my opinion, the most technically elegant and attractive publication of the entire series by Whittington and his associates.) Consider the three main operations:

1. *Excavation and dissection.* If Walcott had been right, all anatomy would be compressed into a single film, and the task of reconstruction would be akin to reviving a cartoon character squashed flat by a steamroller. But what works for Sylvester the cat in a world of fantasy cannot be duplicated for a slab of shale.

Fortunately, the Burgess fossils do not usually lie on a single bedding plane. Engulfed by the mud that buried them, the animals settled into their tombs at various orientations. The mud often infiltrated and sorted their parts into different microlayers, separated by thin veils of sediment—carapace above gills, and gills above legs—thus preserving some three-dimensional structure even when the muds became compressed later on.

By using small chisels or a very fine vibro-drill, not much different from

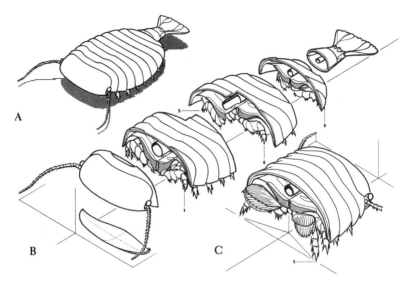

3.3. Reconstruction of *Sidneyia* from a three-dimensional model built in sections by Bruton. (A) The entire animal. (B) The model in six segments, starting from bottom left—the head with its ventral covering plate below, the body in three sections, and the tail piece. (C) The head and front part of the body connected, with the head in the background and to the right. Note the biramous appendages with their walking legs below and gill branches above.

the model in your dentist's office, upper layers can be carefully removed to reveal internal parts beneath. (As these layers are often but microns thick, this delicate work can also be done by hand and with needles, grain by grain or flake by flake.)

Some arthropods are fairly flat, but *Sidneyia,* as the reconstruction shows (figure 3.3), possessed considerable relief; its carapace, or outer covering, formed an arched semicylinder over the soft parts beneath.* In some specimens the underlying gills and legs protrude through a broken carapace, for natural compression and fracturing of specimens is extensive. But Bruton found that he had to go digging in order to reveal an anatomical totality. The appendages of many marine arthropods contain two branches

---

*These outer coverings were, of course, harder than the soft organs below. But the carapaces of most Burgess organisms were not mineralized, and therefore not formed of conventional "hard parts" that fossilize easily. These carapaces were rather like the exoskeletons of modern insects, stiffened but not mineralized. "Lightly sclerotized" might be a better term than "soft-bodied," but the potential for conventional fossilization is nearly nil in either case.

(see pages 104–5, in the inset on arthropod anatomy)—an outer branch bearing gills, used for respiration and swimming, and an inner branch, or walking leg, often used in feeding as well. Hence, as you cut through the outer covering over the center of the body, you first encounter the gill branches, then the leg branches. Bruton found that he could begin with a complete outer covering (figure 3.4), and then dissect through to reveal a layer of gills (figure 3.5), followed by a set of walking legs (figure 3.6). (These drawings are all done directly from the fossils themselves, using a camera lucida attached to a binocular microscope.) Bruton described his method in the conventional passive voice of technical monographs:

> Preparation of specimens shows features . . . to occur at successive levels within the rock and these can be revealed by carefully removing one from above the other, or by removing the thin layer of sediment that separates them. . . . The method of approach has been to remove successively first the dorsal exoskeleton . . . to reveal the filaments of the gills, and then those to expose the leg. Adjacent to the midline where the limb is attached, all three

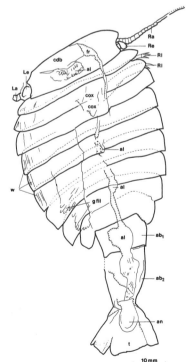

3.4. Camera lucida drawing of a complete specimen of *Sidneyia,* showing the outer covering intact.

3.5. Camera lucida drawing of a *Sidneyia* specimen, primarily showing the gill branches of the appendages underneath the carapace. The incomplete trace of the gut (center) is indicated by oblique stripes. The gill branches are the delicately fingered structures labeled *g* (the number that follows identifies the body segment).

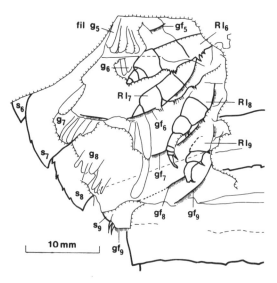

3.6. The walking legs are exposed underneath the gill branches. In this camera lucida drawing, the legs are labeled *Rl,* for "right leg" (the number that follows identifies the body segment).

successive layers, dorsal exoskeleton—gill—leg, lie directly upon each other and it is a matter of hopefully removing an infinitely thin layer of material with the aid of a vibro-chisel (1981, pp. 623–24).

Other rewards lie beneath the outer covering. The alimentary canal runs just beneath the carapace, along the mid-line. One excavated specimen (figure 3.7) revealed a tiny trilobite right in the canal, near the posterior end—a remnant of *Sidneyia*'s last meal before the great mudslide.

2. *Odd orientations.* Since the phyllopod bed was formed by several fossilized mudslides, animals are entombed in a variety of orientations. The majority were buried in their most stable hydrodynamic position, for the mud settled gradually and animals drifted to the bottom. But some came to rest on one side or at an angle—twisted or turned in various ways. In his monograph on the enigmatic *Aysheaia*, Whittington illustrated both the "conventional" orientation, with the animal lying flat, its appendages splayed to the sides, and one of the rarer positions, with the animal twisted and sideways, so that appendages from both sides are compressed and jumbled together (figure 3.8).

Walcott collected specimens in odd orientations, but he tended to ig-

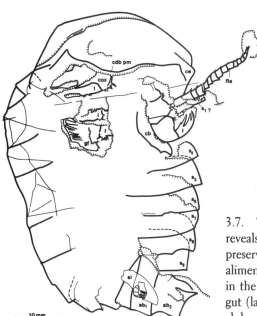

10 mm

3.7. This specimen of *Sidneyia* reveals its last meal, a tiny trilobite preserved in the rear end of the alimentary tract. The trilobite lies in the small exposed portion of the gut (labeled *al*), just above the first abdominal segment (*ab₁*).

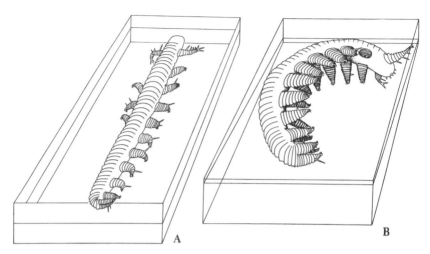

3.8. Two figures from Whittington (1978), illustrating the preservation of *Aysheaia* in various positions. (A) The conventional orientation: we look down on the dorsal, or top, side; the appendages are splayed out in both directions. (B) A much less common orientation: the animal was buried on its side, and the resulting fossil shows one flank, with the appendages of both sides compressed together.

nore them as less informative, and even uninterpretable in their overlapping of different surfaces on a single bedding plane. But Whittington realized that these unusual orientations are indispensable, in concert with specimens in the "standard" position, for working out the full anatomy of an organism. Just as you could not fully reconstruct a house from photos all taken from a single vantage point, "snapshots" at many angles must be combined to reconstruct a Burgess organism. Conway Morris told me that he managed to reconstruct the curious *Wiwaxia*—an animal with no modern relatives, and therefore no known prototype to use as a model—by drawing specimens that had been found in various orientations, and then passing countless hours "rotating the damned thing in my mind" from the position of one drawing to the different angle of another, until every specimen could be moved without contradiction from one stance to the next. Then he finally knew that nothing major was missing or out of place.

Most specimens of *Sidneyia* are preserved in full, flattened view—as if we were looking down from above (as in figure 3.5). This orientation reveals, better than any other, the basic dimensions of body parts, but must leave several questions unresolved, particularly the degree of relief, or rounding, of the body. In this orientation, we can't tell whether *Sidneyia* was a pancake or a tube. Frontal views are needed to reconstruct the basic shape, and to determine some crucial aspects of anatomy not well seen "from above"—the form of the legs in particular.

Figure 3.9, a view from the front, shows the rounded shape of the head, and the positions of insertion for the single pair of antennae and the eyes. Figure 3.10, a head-on view from farther back, illustrates both the rounded shape of the body and a sequence of legs, with their numerous spiny segments all well preserved. We also note the dimensions of the central food groove, running between the coxae, the first segments of the legs, on each side. The gnathobases, the spiny edges of the coxae, border the food groove and give us some appreciation for the probable predatory or scavenging habits of this largest Burgess arthropod. We must assume that large pieces of food were passed forward to the mouth—no wimpy filtrate for this creature. Figure 3.11 shows a close-up of a walking leg, also in frontal orientation.

3. *Part–counterpart.* When you split a rock to find a fossil, you get two for the price of one—the fossil itself (called the part) and the impression of the organism forced into layers above (called the counterpart)—thumb and thumbprint, if you will. The part, as the actual fossil, has been favored by scientists and collectors; the counterpart, as an impression, has less to offer in traditional evaluations. Walcott worked almost exclusively with parts, and frequently didn't bother to keep the counterparts at all. (When he did collect counterparts, he often didn't catalogue them with the matching parts. They ended up in different drawers or relegated to the spoil heaps of less interesting material. Some he even gave away in trade with other museums.)

For a traditional fossil, coherently made of a single piece—the shell of a clam or snail, for example—the distinction between part and counterpart is obvious. The specimen is the part; the mold on the upper surface, the counterpart. Under Walcott's view of Burgess organisms as single films, the same clear difference applies—the film itself is the part; its impression, the less interesting counterpart.

But when Whittington revealed the three-dimensional nature of the Burgess fossils, this easy distinction and differential rating disappeared. An arthropod contains hundreds of articulating pieces; since these are preserved on several adjacent layers in the Burgess Shale, splitting a rock at a bedding plane cannot yield a clear division, with the entire organism (the part) on one surface, and only the impression (the counterpart) on the other. Any split must leave some pieces of the organism on one side, other bits on the opposite block. In fact, the distinction between part and counterpart ultimately breaks down for the Burgess fossils. You can only say that one surface preserves more interesting anatomy than the other. (By convention, the Burgess workers finally decided to designate the top view upon the organism as the part, and the view looking up as the counterpart. By this scheme, for an animal like *Sidneyia,* eyes, antennae, and other features of the external carapace are often preserved on the counterpart, legs and internal anatomy on the part.)

All expeditions from 1966 to the present have rigorously collected both part and counterpart (when preserved), keeping and cataloguing them together. Some of the greatest Burgess discoveries of the past twenty years have occurred at the Smithsonian when a Walcott counterpart, sometimes uncatalogued, sometimes even classified in a different phylum, was recognized and reunited with its part. Can you top this for a heart-warming tale, more satisfying (since less probable) than the reunion of Gabriel with Evangeline? In 1930, the Raymond expedition found a specimen of *Branchiocaris pretiosa,* an exceedingly rare arthropod with fewer than ten known examples. In 1975 (when Derek Briggs had already submitted his monograph on this species for publication), the Royal Ontario Museum expedition found the counterpart of this specimen, still lying on the talus slope in British Columbia where Raymond and his party had spurned it forty-five years before!

Obviously, if both part and counterpart contain important bits of anatomy, we must study them together if we strive for tolerable completeness

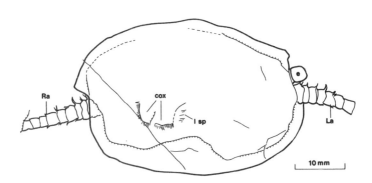

3.9. Camera lucida drawing of a specimen of *Sidneyia* preserved in an unusual orientation. We are looking at the front end head on, and therefore can appreciate the convexity of the animal—information that we cannot get in the usual orientation. Note in particular the positions of insertion for the antennae (labeled *Ra* and *La*) and for the eye *(e).*

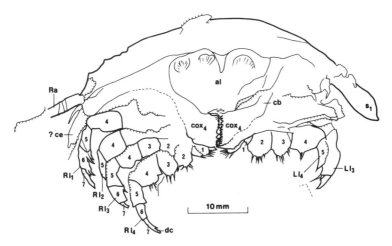

3.10. A specimen of *Sidneyia* in an unusual orientation that reveals the arrangement of the legs. We are looking head on at a cross section through the front end of the body, just behind the head, and can see the first four legs on the animal's right side, compressed together (labeled Rl_1–Rl_4). The alimentary canal *(al)*, in the center of the body, is also visible.

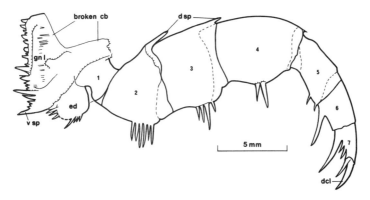

3.11. Camera lucida drawing of a walking leg of *Sidneyia*. Note the strong spines (labeled *gn*, for "gnathobase") at the point of insertion for the leg into the body. This array of spines bordering the food groove suggests that the animal was a predator. The leg is so well preserved that we can count the segments and infer the orientation in life.

in reconstruction. (In their camera lucida drawings, Whittington and colleagues have followed the convention of including information from both part and counterpart in the same figure.) Reassociation of part with counterpart has resolved a puzzle in the study of *Sidneyia*. Based on an isolated specimen, Walcott had suggested a peculiar reconstruction for the gills of *Sidneyia*. But Bruton, examining both Walcott's part and the "counterpart which Dr. D. E. G. Briggs observantly found among uncatalogued material in the Walcott Collection" (Bruton, 1981, p. 640), discovered that the supposed gill did not belong to *Sidneyia* at all. Conway Morris later identified this fossil as a decayed and folded specimen of the priapulid worm *Ottoia prolifica*.

These three procedures—excavation, odd orientations, and part–counterpart—are guides to the three dimensional reanimation of squashed and distorted fossils. They don't tell us much about other aspects of life among Burgess organisms—how they moved and ate, and how they grew, for example. Unfortunately, for all its virtues in preserving anatomy, the Burgess Shale, as a transported assemblage buried in a mud cloud, does not provide other kinds of evidence that more conventional faunas often include. We have no tracks or trails, no burrows, no organisms caught in the act of eating their fellows—in short, few signs of organic activity in process. For some reason not well understood (and most unfortunately), the Burgess Shale includes almost no juvenile stages of organisms.

Still, some procedures beyond those already noted have been useful in particular cases; they will be discussed in turn as the organisms enter our story. I have already mentioned the gut contents of *Sidneyia*. Other organisms have also been identified as carnivores by a study of their alimentary tracts. For example, in the gut of a priapulid worm Conway Morris found smaller members of the same species—the world's earliest example of cannibalism—and numerous hyolithids. He also used varying degrees of decay to resolve the anatomy of the priapulid worm *Ottoia prolifica*. Bruton (for *Sidneyia, Leanchoilia,* and *Emeraldella*) and Briggs (for *Odaraia*) made three-dimensional models from composites of their drawings and photographs. Conway Morris has used injuries and patterns of growth to understand the habits of the enigmatic *Wiwaxia*. He argues (1985) that in a unique Burgess example of growth caught in the act, one specimen was buried in the process of molting—casting off an old garment for an entirely new outer coat of plates and spines.

THE CHRONOLOGY OF
A TRANSFORMATION

What do scientists "do" with something like the Burgess Shale, once they have been fortunate enough to make such an outstanding discovery? They must first perform some basic chores to establish context—geological setting (age, environment, geography), mode of preservation, inventory of content. Beyond these preliminaries, since diversity is nature's principal theme, anatomical description and taxonomic placement become the primary tasks of paleontology. Evolution produces a branching array organized as a tree of life, and our classifications reflect this genealogical order. Taxonomy is therefore the expression of evolutionary arrangement. The traditional medium for such an effort is a monograph—a descriptive paper, with photographs, drawings, and a formal taxonomic designation. Monographs are almost always too long for publication in traditional journals; museums, universities, and scientific societies have therefore established special series for these works. (As noted before, most Burgess descriptions have appeared in monographs published by the Royal Society of London in their *Philosophical Transactions*—a series for long papers.) These monographs are expensive to produce and have strictly limited circulation, mostly to libraries.

This situation has engendered the unfortunate condescension expressed toward monographs and their authors by many scientists from other disciplines. These works are dismissed as exercises in "mere description," a kind of cataloguing that could as well be done by clerks and drones. At most, some credit may be given for care and attention to detail—but monographs do not emerge as the vanguard of creative novelty.

Some monographs are pedestrian, of course—the description of a new brachiopod or two from a well-known formation deposited during the heyday of the group's success will raise few eyebrows—but then a great deal of workaday physics and chemistry is also dial-twirling to iterate the obvious. The best monographs are works of genius that can transform our views about subjects inspiring our passionate interest. How do we know about Lucy, the "ape-man of Java," our Neanderthal cousins, the old man of Cro-Magnon, or any of the other human fossils that fire our imagination as fully as an Apollo landing on the moon, except by taxonomic monographs? (Of course, in these cases of acknowledged "newsworthiness," highly touted preliminary reports long precede any technical publication, usually providing, as the cliché goes, much heat with little light.)

The world is so full of a number of things,
I'm sure we should all be as happy as kings.
                              —Robert Louis Stevenson

This famous couplet, from A Child's Garden of Verses, *expresses the chief delight of our natural world and the primary result of evolution—incredible and irreducible variety. Since the human mind (in its adult version, at least) craves order, we make sense of this variety by systems of classification. Taxonomy (the science of classification) is often undervalued as a glorified form of filing—with each species in its folder, like a stamp in its prescribed place in an album; but taxonomy is a fundamental and dynamic science, dedicated to exploring the causes of relationships and similarities among organisms. Classifications are theories about the basis of natural order, not dull catalogues compiled only to avoid chaos.*

Since evolution is the source of order and relationship among organisms, we want our classifications to embody the cause that makes them necessary. Hierarchical classifications work well in support of this aim because the primary topology of life's tree—the joining of twigs to branches, branches to limbs, and limbs to trunks as we trace species back to ever earlier common ancestors—can be expressed by a system of ever more inclusive categories. (People join with apes and monkeys to make primates; primates with dogs to make mammals; mammals with reptiles to make vertebrates; vertebrates with insects to make animals, and so on. Since Linnaeus and other pre-Darwinians also used hierarchical systems, evolution is not the only possible source of order expressed by this form; but evolution by diversification does imply branching from common ancestry, and such a topology is best rendered by hierarchical classification.)

Modern taxonomies recognize seven basic levels of increasing inclusion—from species (considered as the fundamental and irreducible units of evolution) to kingdoms (the broadest groupings of all): species, genera, families, orders, classes, phyla, and kingdoms.

At the highest level—the kingdom—the old folk division into plants and animals, and the old schoolboy system of plants, animals, and single-celled protists, have been largely superseded by a more convenient and accurate five-kingdom system: Plantae, Animalia, and Fungi for multicellular organisms; Protista (or Protoctista) for single-celled organisms with complex cells; and Monera for single-celled organisms (bacteria and cyanophytes) with simple cells devoid of nuclei, mitochondria, and other organelles.

The next level—the phylum—is the basic unit of differentiation within kingdoms. Phyla represent the fundamental ground plans of anatomy. Among animals, for example, the broadest of basic groups are designated as phyla—sponges, "corals" (including hydras and jellyfish), annelids (earthworms, leeches, and marine polychaetes), arthropods (insects, spiders, lobsters, and the like), mollusks (clams, snails, squid), echinoderms (starfishes, sea urchins, and sand dollars), and chordates (vertebrates and their kin). In other words, phyla represent the major trunks of life's tree.

This book treats the early history of the animal kingdom. In focusing on the origin of phyla and their early number and degree of differentiation, we ask the most basic of all questions about the organization of our animal kingdom.

How many phyla of animals does our modern earth contain? Answers vary, since this question involves some subjective elements (a terminal twig is an objective thing, and species are real units in nature, but when is a branch large enough to be called a bough?). Still, we note some measure of agreement; phyla tend to be big and distinct. Most textbooks recognize between twenty and thirty animal phyla. Our best modern compendium, a book explicitly dedicated to the designation and description of phyla (Margulis and Schwartz, 1982) lists thirty-two animal phyla—a generous estimate in comparison with most. In addition to the seven familiar groups already mentioned, the animal phyla include, among others, the Ctenophora (comb jellies), Platyhelminthes (flatworms, including the familiar laboratory Planaria), Brachiopoda (bivalved invertebrates common as Paleozoic fossils, but rarer today), and Nematoda (unsegmented roundworms, usually tiny and fantastically abundant in soil and as parasites).

After such a long disquisition, the point of this exegesis with respect to the Burgess Shale may be quickly stated: the Burgess Shale, one small quarry in British Columbia, contains the remains of some fifteen to twenty organisms so different one from the other, and so unlike anything now living, that each ought to rank as a separate phylum. We hesitate to give such a "high" designation to single species because our traditions dictate that phyla achieve their distinctness through hundreds of speciation events, each building a bit of the total difference, piece by piece. Hence, the anatomy of a group should not become sufficiently distinct to rank as a separate phylum until a great deal of diversity has been accumulated by repeated speciation. According to this conventional view—obviously incorrect or incomplete by evidence from the Burgess—lineages of one or a few species cannot diverge far enough to *rank as phyla. But* que faire? *The fifteen to twenty unique Burgess designs are phyla by virtue of anatomical uniqueness. This remarkable fact must be acknowledged with all its implications, whatever decision we ultimately make about the formalities of naming.*

The worst of human narrowness pours forth in the negative assessment of monographic work as merely descriptive. Scientific genius is equated with an oddly limited subset of intellectual activities, primarily analytical ability and quantitative skill, as though anyone could describe a fossil but only the greatest thinkers could conceive of the inverse-square law. I wonder if we will ever get past the worst legacy of IQ theory in its unilinear and hereditarian interpretation—the idea that intelligence can be captured by a single number, and that people can be arrayed in a simple sequence from idiot to Einstein.

Genius has as many components as the mind itself. The reconstruction of a Burgess organism is about as far from "simple" or "mere" description as Caruso from Joe Blow in the shower, or Wade Boggs from Marvelous Marv Throneberry. You can't just look at a dark blob on a slab of Burgess shale and then by mindless copying render it as a complex, working arthropod, as one might transcribe a list of figures from a cash-register tape into an account book. I can't imagine an activity further from simple description than the reanimation of a Burgess organism. You start with a squashed and horribly distorted mess and finish with a composite figure of a plausible living organism.

This activity requires visual, or spatial, genius of an uncommon and particular sort. I can understand how this work proceeds, but I could never do it myself—and I am therefore relegated to *writing* about the Burgess Shale. The ability to reconstruct three-dimensional form from flattened squashes, to integrate a score of specimens in differing orientations into a single entity, to marry disparate pieces on parts and counterparts into a functional whole—these are rare and precious skills. Why do we downgrade such integrative and qualitative ability, while we exalt analytical and quantitative achievement? Is one better, harder, more important than the other?

Scientists learn their limitations and know when they need to collaborate. We do not all have the ability to assemble wholes from pieces. I once spent a week in the field with Richard Leakey, and I could sense both his frustration and his pride that his wife Meave and their coworker Alan Walker could take tiny fragments of bone and, like a three-dimensional jigsaw puzzle, put together a skull, while he could do the work only imperfectly (and I saw nothing at all but fragments in a box). Both Meave and Alan showed these skills from an early age, largely through a passion for jigsaw puzzles (curiously, both, as children, liked to do puzzles upside down, working by shapes alone, with no help from the picture).

Harry Whittington, who shares this rare visual genius, also expressed his gift at an early age. Harry began with no particular advantages of class or culture. He grew up in Birmingham, the son of a gunsmith (who died when Harry was only two) and grandson of a tailor (who then raised him). His interests wandered toward geology, thanks largely to the inspiration of a sixth-form (just pre-university) geography teacher. Yet Harry had always recognized and exploited his skill in three-dimensional visualization. As a child, he loved to build models, mostly of cars and airplanes, and his favorite toy was his Meccano set (the British version of an Erector set, providing strips of steel that can be bolted together into a variety of structures). In beginning geology courses, he excelled in map interpretation and, especially, in drawing block diagrams. The consistent theme is unmistakable: a knack for making three-dimensional structures from two-dimensional components, and inversely, for depicting solid objects in plane view. This ability to move from two to three dimensions, and back again, provided the key for reconstructing the fauna of the Burgess Shale.

Harry Whittington was clearly the best possible person for the Burgess project. He was not only the world's leading expert on fossil trilobites (the most conspicuous arthropods of the fossil record), but he had done his most elegant work (Whittington and Evitt, 1953, for example) on rare three-dimensional specimens preserved in silica. The original calcium carbonate of these fossils had been replaced by silica, while the surrounding limestone retained its carbonate base. Since carbonates are dissolved by hydrochloric acid, while silicates are unaffected, the matrix could be dissolved away, providing the rare advantage of three-dimensional preservation completely separable from the surrounding rock. Whittington had therefore been blessed with an ideal, if unwitting, preparation for the Burgess Shale many years later. He had studied three-dimensional structure within rock and then been able to judge his hunches and hypotheses by dissolving the matrix and recovering the fossils intact. These studies "preadapted" Whittington, to use a favorite word in the jargon of evolutionary biology, for his discovery and exploitation of three-dimensional structure in the Burgess Shale fossils.

Don't accept the chauvinistic tradition that labels our era the age of mammals. This is the age of arthropods. They outnumber us by any criterion—by species, by individuals, by prospects for evolutionary continuation. Some 80 percent of all named animal species are arthropods, the vast majority insects.

The higher-level taxonomy of arthropods therefore becomes a subject of much concern and importance. Many schemes have been proposed, and their differences continue to inspire debate. But general agreement can be wrested from most quarters concerning the number and composition of basic subgroups within the phylum. (The evolutionary relationships among

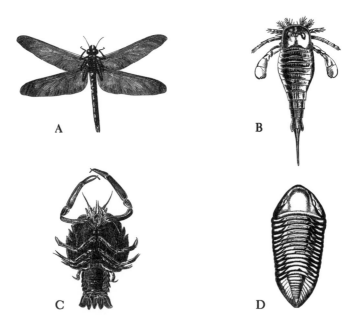

1. Representative fossil specimens of the four great groups of arthropods, taken from the most widely used textbook in the history of paleontology, the late-nineteenth-century work of Zittel. (A) A giant dragonfly from the Carboniferous, representing the Uniramia. (B) A fossil eurypterid, representing the Chelicerata. The first pair of head appendages is small and hidden under the carapace; the other five pairs are visible in this figure. (C) A fossil crab, representing the Crustacea. (D) A trilobite.

subgroups are more problematical, but this subject will not be a major concern of this book).

The scheme that I follow here is conservative and traditional, the closest to consensus that can be achieved. I recognize four major groups, three still living, one exclusively fossil (figure 1), and I make no proposal about evolutionary connections among them.

1. Uniramia, including insects, millipedes, centipedes, and perhaps also the onychophores (a small and unusual, but particularly fascinating group, of which a good deal more later on, for the Burgess Shale contains a probable member).

2. Chelicerata, including spiders, mites, scorpions, horseshoe crabs, and the extinct eurypterids.

3. Crustacea, primarily marine (the terrestrial pillbug, an isopod, ranks as an exception), and including several groups of small bivalved forms, little known to nonprofessionals, but fantastically diverse and common in the oceans (copepods and ostracodes, for example), the barnacles, and the decapods (crabs, lobsters, and shrimp), whom we eat with relish while regarding their insect cousins as disgusting and unpalatable.

4. Trilobita, everybody's favorite invertebrate fossil, extinct for 225 million years, but common in Paleozoic rocks.

Since the resolution of the Burgess Shale fauna depends so centrally upon an understanding of the amazingly diverse and disparate arthropods, we must enter into some details of arthropod anatomy. Lest this prospect sound daunting, let me assure you that I shall keep the jargon to an absolute and fully comprehensible minimum—only about twenty terms from among more than a thousand available. (I shall not list these terms, but rather define them in the course of discussion. All key terms are underlined at their first use.)

The basic principle of arthropod design is metamerism, the construction of the body from an extended series of repeated segments. The key to arthropod diversification lies in recognizing that an initial form composed of numerous nearly identical segments can evolve by reduction and fusion of segments, and by specialization of initially similar parts on different segments, into the vast array of divergent anatomies seen in advanced arthropods. Fortunately, we can grasp the complexities of this central theme in arthropod evolution by considering just two matters: the fusion and differentiation of segments themselves, and the specialization of appendages.

The numerous separate and similar segments of ancestral arthropods (figure 2) tended to coalesce into fewer specialized groups. The most common arrangement is a three-part division, into head, middle, and rear (called by various names, such as cephalon, thorax, and pygidium in trilobites, or head, thorax, and abdomen in insects and crustaceans). Most chelicerates have a two-part division, with a _prosoma_ followed by an _opisthosoma_. The fused tailpiece of many crustaceans is called a _telson_.

Arthropods have external skeletons, or _exoskeletons_ (stiff, but unmineralized in most groups, thus explaining the rarity of many arthropods as fossils). As segments fused, their exoskeletal parts joined to form discrete skeletal units called _tagma_. This process of fusion is called _tagmosis_. Different patterns of skeletal tagmosis provide a primary criterion for identifying fossil arthropods.

Just as important, and as crucial to the Burgess story, is the specialization and differentiation of appendages. Each segment of the original, unspecialized, many-segmented arthropod bore a pair of appendages—one on each side of the body. Each appendage consisted of two branches, or _rami_ (singular _ramus_). These rami are named according to their position—the _inner ramus_ and the _outer ramus_—or according to their usual function. Since the outer branch often bears a gill used in respiration or swimming (or both), it is often called the _gill branch_. The inner branch is usually used in locomotion, and may be called the _leg branch_, _walking branch_, or _walking leg_. (The common term "walking leg" may strike readers as amusingly redundant, but "leg" is an anatomical, not a functional term, and not all arthropods use their legs for walking; insect mouth parts, for example, are slightly modified legs.)

This original structure (figure 3) is called a _biramous_ (literally, "two-branched") limb. (If you retain no other term from this discussion,

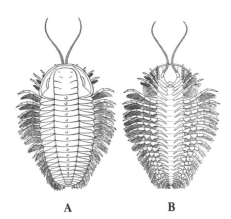

A   **B**

2. The numerous similar segments of a primitive arthropod, as seen in the trilobite _Triarthrus_. With the exception of the frontal antennae, all pairs of appendages are similar and biramous, and each body segment has a single pair. (A) Top view. (B) Bottom view. From Zittel.

104

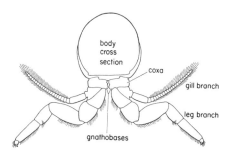

3. Cross section through a body segment of an arthropod, showing a pair of typical biramous limbs. Drawn by Laszlo Meszoly.

please inscribe the definition of a biramous limb in your long-term memory. It is the single most important facet of arthropod anatomy in our Burgess discussion.) Specialized arthropods often lose one of the two branches, retaining the other as a <u>uniramous</u> ("one-branched") limb. (Please place "uniramous" next to "biramous" in your long-term memory.) The higher-level taxonomy of arthropods records the different mixes of uniramous and biramous limbs on various parts of the body.

The walking legs of most marine arthropods perform an additional function that seems odd from our vertebrate-centered perspective. Some marine arthropods feed as we do by seizing food items in front of their head and passing them directly to the mouth. But most use their walking legs to grasp food particles and pass them forward to the mouth along a <u>food groove</u> situated in the <u>ventral</u> (bottom) mid-line, between the legs. (The top side of an animal is called <u>dorsal</u>.) <u>Arthropod</u> means "jointed foot," and the appendages are composed of several segments. Segments located near the body are <u>proximal</u>; those far away at the ends of the appendage are <u>distal</u>. The most proximal segment of the walking leg is called a <u>coxa</u>. The edge of the coxa bordering the food groove is often armed with teeth, used to capture and move the food forward (see figure 3) and called a <u>gnathobase</u> (literally, "jawed foundation").

We form the higher-level taxonomy of arthropods by joining the two principles discussed above: patterns of tagmosis, or fusion of segments, and specialization of appendages by loss of one ramus and differentiation of the other. Beginning with an ancestral arthropod built of many unfused segments, each bearing a biramous limb, the major groups have evolved along different routes of tagmosis and specialization. Consider the four major kinds of arthropods:

1. Uniramia. As the name implies, insects and their kin have invariably lost the gill branch of the original biramous limb; they build their appendages (antennae, legs, mouth parts) exclusively from leg branches. (Insects breathe through invaginations of the external body surface, called tracheae.)

2. Chelicerata. Most modern chelicerates have six uniramous appendages on the prosoma. The first pair—chelicerae—are jawlike at the distal end and are used for grasping. (Antennae are absent in this group.) The second pair—pedipalps—are usually sensory in function. The last four pairs are usually leglike (giving spiders their eight legs). All these anterior appendages have evolved from leg branches. The situation is reversed on the posterior section. The opisthosomal appendages are also uniramous, but have been built from gill branches only. (The "lung-books," or breathing organs, of spiders are on the abdomen.)

3. Crustacea. Despite an enormous diversity of form, from barnacles to lobsters, all crustaceans are distinguished by their stereotypical pattern of five pairs of appendages on the head (indicating that the head was formed by a tagmosis of at least five segments). The first two pairs, usually called antennae and antennules, are uniramous; they lie in a pre-oral position, in front of the mouth, and have sensory functions. The last three lie in a post-oral position, behind the mouth, and are usually used in feeding, as mouth parts. Appendages on the trunk often retain the original biramous form.

4. Trilobita. The trilobite head bears one pre-oral pair of appendages (antennae) and three post-oral pairs. Each body segment usually bears a pair of biramous limbs very little modified from the presumed ancestral form.

The stereotypy of these patterns is, perhaps, the most striking phenomenon in modern arthropods. Of nearly a million described species of insects, none has a biramous appendage, and nearly all have exactly three pairs of limbs on the thorax. Marine crustacea display incredible diversity of form, but all have the same pattern of tagmosis in the head—two pre-oral and three post-oral pairs of appendages. Apparently, evolution settled upon just a few themes or ground plans for arthropods and then stuck with them through the greatest story of diversification in the entire animal kingdom.

The story of the Burgess Shale ranks as perhaps the most amazing in the history of life largely in relation to this phenomenon of later restriction in arthropod ground plans—for in addition to early representatives of all four later groups, the Burgess Shale, one quarry in British Columbia, contains fossils of more than twenty additional basic arthropod designs. How could such disparity originate so quickly? Why did only four basic designs survive? These questions form the primary subject of this book.

If Harry Whittington had known at the outset what a restudy of the Burgess Shale would require in time and commitment, he would probably not have started. He was fifty years old during the first field season of 1966, and already had enough commitments to last a lifetime. Moreover, as professor of geology at Cambridge he had oppressive administrative responsibilities that could not be delegated.

But the Burgess was too beautiful and variegated a plum to resist. Besides, everybody knew that its arthropods—the focus of Whittington's proposed work—posed no major taxonomic dilemmas. Harry told me that when he first decided to work on the Burgess, he "expected to spend a year or two describing some arthropods—full stop." In England, a "full stop" is a period—ending the sentence, and ending the project.

It was not to be. Harry Whittington spent four and a half years just writing his first monograph on the genus *Marrella*. Surprise cascaded upon surprise, starting slowly with doubts about the identity of certain arthropods, and accelerating until a new interpretation jelled in the mid-1970s. This view blossomed to guide all subsequent work toward a new conception for the history of early life. As I read the taxonomic monographs in chronological order, I came to see this story as a classical drama in five acts. No one was killed; few people even got angry. But just as Darwin let his theory gestate for twenty-one basically quiet years between formulation and publication, the similar time for the reevaluation of the Burgess Shale has produced, behind a placid exterior, an intellectual drama of the highest order.

The Burgess Drama

ACT 1. *Marrella* and *Yohoia*: The Dawning and Consolidation of Suspicion, 1971–1974

THE CONCEPTUAL WORLD THAT WHITTINGTON FACED

Harry Whittington is, by nature, a cautious and conservative man. To this day, though he served as midwife to a major transformation of thought, he views himself as an empiricist, with skill in the meticulous description of arthropod fossils. His favorite motto exhorts his younger colleagues to place fact and description before theory, for "one should not run before one can walk."

Whittington began, as would any paleontologist who believes in cranking up slowly and deliberately, with the genus *Marrella,* the most common organism in the Burgess Shale. *Marrella splendens* overwhelms anything else in the Burgess by sheer abundance. Walcott collected more than 12,000 specimens. Whittington's party gathered another 800, and I am custodian to 200 more, collected by Percy Raymond in 1930. Many Burgess species are known from fewer than ten specimens, some from only one. But with nearly 13,000 potential views, one need hardly worry about destroying unique evidence by dissection, or failing to find a crucial orientation.

Marrella splendens is the first Burgess organism that Walcott found and drew; it virtually identifies the Burgess Shale. When Walcott described *Marrella* formally in 1912, he recognized that his "lace crab" was not a conventional trilobite, but still placed *Marrella* in the class Trilobita, order previously unknown. Following his need to view Burgess organisms as primitive members of later successful groups, he wrote: "In *Marrella* the trilobite is foreshadowed" (1912, p. 163).

Not all of Walcott's colleagues were convinced. In the Smithsonian archives, I found some interesting correspondence with Charles Schuchert, celebrated Yale paleontologist and codifier of the canonical legend about Walcott's discovery of the Burgess Shale. After reading his friend Walcott's paper on the Burgess arthropods, Schuchert wrote to him on March 26, 1912:

> To you personally I want to say that from the first time that I saw *Marrella* and now with your many excellent pictures of this animal I still cannot get it into my head that this is a trilobite. . . . I cannot see how it can be a trilobite. Such gills are unknown, I believe, in any trilobite. However, I am only throwing out these half-digested ideas for your consideration rather than to convince you that *Marrella* is not a trilobite.

Yet Schuchert, as committed as Walcott to the larger theme that all Burgess creatures belong in known groups, never suggested uniqueness for *Marrella*, but only hinted at a different home among well-known arthropods.

To give some idea of the conceptual barriers that Whittington faced when he began to redescribe the arthropods of the Burgess Shale, I must now exemplify what I shall call, throughout this volume, "Walcott's shoehorn"—his decision to place all Burgess genera in established major groups. Most readers will need to consider these pages in conjunction with the insets on taxonomy and arthropod anatomy (pages 98 and 102). I am asking some investment here from readers with little knowledge of invertebrate biology. But the story is not difficult to follow, the conceptual rewards are great, and I shall try my best to provide the necessary background and guidance. The material is not at all conceptually difficult, and the details are both beautiful and fascinating. Moreover, you can easily retain the thread of argument without completely following the intricacies of classification—as long as you realize that Walcott and all students of the Burgess before Whittington placed these organisms in conventional groups, and that Whittington slowly weaned himself away from this tradition, and toward a radical view about the history of life's diversification.

Walcott presented his complete classification of Burgess arthropods on page 154 of his 1912 paper (reproduced here as table 3.1). He scattered his Burgess genera widely among four subclasses, all placed within his version of the class Crustacea. Walcott defined Crustacea far more broadly than we do today. He included virtually all marine and freshwater arthropods, organisms that span the entire arthropod phylum as defined today. Of his four subclasses, the modern branchiopods (1) are a group of predominantly freshwater crustaceans, including the brine shrimp and the cladocerans, or water fleas; malacostracans (2) form the great group of marine crustacea, including crabs, shrimp, and lobsters; trilobites (3) are, of course, the most famous of fossil arthropods; while merostomes (4), including fossil euryp-terids and modern horseshoe crabs, are closely related to terrestrial scorpions, mites, and spiders.

The fate of Walcott's 1912 chart is a striking epitome of the entire Burgess story. Of his twenty-two genera, only two are legitimate members of their groups. *Nathorstia* (now called *Olenoides serratus*) is an uncontroversial trilobite (Whittington, 1975b); *Hymenocaris* (now called *Canadaspis*) is a true crustacean of the malacostracan line (see Act 3). Three genera *(Hurdia, Tuzoia,* and *Carnarvonia)* are bivalved arthropod carapaces with no soft parts preserved; they cannot be properly allocated to any arthropod subgroup, and remain unclassified today. Three other names do not belong to the story of Burgess arthropods: *Tontoia,* position still unresolved and possibly inorganic, comes from the Grand Canyon, not the Burgess Shale; *Bidentia* is an invalid name, and these specimens belong to the genus *Leanchoilia; Fieldia,* misidentified by Walcott, is a priapulid worm, not an arthropod.

Of the remaining fourteen genera, two *(Opabinia* and *Anomalocaris)* have been reallocated to unique phyla bearing no known relationship to modern groups; they, and at least a dozen others of similar status (classified, for the most part, as annelid worms by Walcott), form the centerpiece of my story. Another eleven have been taken from the known and comfortable homes that Walcott designated, and reclassified as arthropods of unique anatomy, outside the range of any other modern or fossil group.

TABLE 3.1.   *Walcott's 1912 Classification of Burgess Arthropoda*

Crustacea   class

  1. Branchiopoda   subclass

   Anostraca   order

   Opabinia
   Leanchoilia
   Yohoia
   Bidentia

   Notostraca   order

   Naraoia
   Burgessia
   Anomalocaris
   Waptia

  2. Malacostraca   subclass

   Hymenocaris [*Canadaspis*]
   Hurdia
   Tuzoia
   Odaraia
   Fieldia
   Carnarvonia

  3. Trilobita   subclass

   Marrella
   Nathorstia [*Olenoides serratus*]
   Mollisonia
   Tontoia

  4. Merostomata   subclass

   Molaria
   Habelia
   Emeraldella
   Sidneyia

Only *Naraoia*, which Walcott classified as a branchiopod crustacean, belongs in a known group, though Walcott chose the wrong one. *Naraoia* is, in fact, a highly peculiar trilobite (Whittington, 1977).

When I state that no one challenged Walcott's shoehorn until Whittington and colleagues redescribed the Burgess Shale, I do not mean that all paleontologists accepted Walcott's specific allocations. Articles on Burgess organisms were sparse during the sixty years between Walcott's descriptions and Whittington's first monograph—especially considering the importance of the fauna, as acknowledged by all paleontologists*—but the limited literature proposed several schemes for taxonomies departing strongly from Walcott's.

But these alternatives, however varied among themselves, never abandoned a strict allegiance to Walcott's larger presupposition—the shared, and almost always unstated, view of paleontologists that fossils fall into a limited number of large and well-known groups, and that life's history generally moves toward increasing complexity and diversity.

Leif Størmer drew the task of describing most Burgess Shale arthropods for the collectively written *Treatise on Invertebrate Paleontology,* and published his results (Størmer, 1959) in a large volume devoted primarily to trilobites. Størmer's solution was diametrically opposed to Walcott's. Instead of spreading the Burgess arthropods widely among groups throughout the phylum, he brought most of them together in allegiance with the trilobites themselves. He could not, of course, claim that all these diverse

---

*I asked Whittington why so little work had been done before his redescriptions, for Walcott's specimens had always been available at the Smithsonian. He cited a number of reasons, all no doubt contributory, but not enough in their ensemble to explain this curious fact. Walcott's wife, for one, was quite possessive and discouraging, though she held no proprietary power over the specimens. She hated Percy Raymond for collecting again at the Burgess so soon after her husband's death in 1927. Raymond, in his turn, had been no fan of Walcott's, and taunted him as "the great executive paleontologist" for letting administrative work absorb all his time, thus precluding a proper study of the Burgess fossils. (This was an unusually acerbic assessment for Raymond, who was the most mild-mannered of men. Al Romer, who knew him well, once told me that Raymond was at the bottom of a familial pecking order, with his wife, children, and dog above him. His favorite hobby, collecting pewterware, definitely contributed to his non-macho image.) While Walcott lived, no one else would work on the specimens, for he always intended to do a proper study himself, and no one dared upstage the most powerful man in American science. (Such proprietary claims are traditionally honored in paleontology, even for scientists low on the totem pole; discovery implies the right of description, with a statute of limitation often construed as extending for a lifetime.) Walcott's wife, and the memory of his power, managed to extend a reluctance for work on Burgess material even beyond Walcott's grave. Moreover, as Whittington reports, although the "type" specimens were accessible (the few used in the original descriptions of the species), almost all the material resided in drawers placed high in cabinets, and therefore unavailable for casual browsing—the serendipitous mode of origin for many paleontological studies. They also were housed in a building without air conditioning (now remedied). Most paleontologists work in universities, and have substantial free time only during the summer. Need I say more to anyone who has experienced the pleasures of our nation's capital in July or August!

and most untrilobite-like animals *truly* belonged to the class Trilobita proper. But he did neatly (however falsely) resolve the problem of arthropod disparity in the Burgess by placing all the major genera in one supposedly coherent evolutionary group, lying right next to the Trilobita. He called his group the Trilobitoidea (literally, "trilobite-like").

This solution may seem too pat or arbitrary to be believed. But Størmer had a rationale (invalidated, as we shall see, by later advances in taxonomic theory). He acknowledged, of course, the great range of form among Burgess arthropods, but he forged a taxonomic union because he argued that they all possessed the same kind of "primitive" appendages on body segments behind the head—a biramous, or two-pronged, form with a gill branch above a leg branch (see inset, page 104). Since trilobites also possessed appendages of this form, the Trilobita proper and the Trilobitoidea (the heterogeneous Burgess oddballs) could be grouped together in a larger taxon, called Trilobitomorpha. Størmer presented the following rationale:

> The Trilobitomorpha are linked together by the seemingly common basic structure of their appendages. Since the trilobite limb appears to be a characteristic and conservative structure, its presence in fossil arthropods may be interpreted as evidence of close relationship between the many different forms possessing it (1959, p. 27).

Størmer's classification of the Trilobitoidea is shown in table 3.2. All but two of his sixteen genera reside exclusively in the Burgess Shale (*Tontoia*, as previously mentioned, comes from the Grand Canyon; *Cheloniellon*, from the Devonian *Lagerstätte* of the Hunsrückschiefer). Størmer divided the Burgess genera into three groups: (1) *Marrella* alone; (2) the cluster that Walcott had aligned with the merostomes, or horseshoe-crab group, a superficial similarity that Størmer acknowledged in his name Merostomoidea ("merostome-like"); (3) the genera that Walcott had placed in the Notostraca, a group of branchiopod crustaceans (a superficial similarity honored by Størmer in his chosen name Pseudonotostraca). Yet, try as he might, Størmer could not comfortably squeeze all the Burgess forms into his Trilobitoidea. Four genera stumped him, and he tacked them onto the end of his classification as "subclass Uncertain"—a solution neither elegant nor Latin.

I have presented this detailed contrast of Størmer's system with Walcott's original scheme for two reasons. First, the power of the shoehorn can be illustrated by demonstrating that all taxonomic solutions, however divergent in a plethora of details, worked within this unchallenged postulate. Both Walcott's scattering into a broad range of known groups, and

Table 3.2. *Størmer's 1959 Classification of Trilobitoidea*

Trilobitomorpha subphylum

  Trilobita class

  Trilobitoidea class

    1. Marrellomorpha   subclass

    Marrella

    2. Merostomoidea   subclass

    Sidneyia
    Amiella
    Emeraldella
    Naraoia
    Molaria
    Habelia
    Leanchoilia

    3. Pseudonotostraca   subclass

    Burgessia
    Waptia

    4. subclass   Uncertain

    Opabinia
    Cheloniellon
    Yohoia
    Helmetia
    Mollisonia
    Tontoia

Størmer's gathering together as the Trilobitoidea remained fully faithful to the rule of the shoehorn—all Burgess genera belong in established groups. Second, Størmer's interpretation, published in the major compendium of international opinion, was the most up-to-date, standard classification of Burgess arthropods when Whittington started his project. Størmer's Trilobitoidea was Whittington's context as he began his monograph on *Marrella*.

Marrella: first doubts

Harry Whittington's initial monograph on *Marrella* (1971) scarcely reads like the stuff of revolution—at first glance. It begins with an introduction by Y. O. Fortier, director of the Geological Survey of Canada. Parroting the traditional assumptions of Walcott's shoehorn and the cone of increasing diversity, Mr. Fortier launched the entire enterprise with the following paragraph:

> The Burgess Shale of Yoho National Park, British Columbia, is world famous and unique. It was from these fossiliferous Cambrian beds that Charles D. Walcott . . . collected and subsequently described . . . a remarkable and diversified group of fossils that represent the *primitive ancestors of nearly every class of arthropod* as well as several other animal Phyla [my italics].

Whittington's title contains no hint of the shape of things to come. He followed the standard form of taxon, place and time—what my former student Warren Allmon calls "*x* from the *y*-ity of *z*-land." He even adopted—but for the only time, and much to his later regret—Størmer's name Trilobitoidea: "Redescription of *Marrella splendens* (Trilobitoidea) from the Burgess Shale, Middle Cambrian, British Columbia."

Marrella is a small and elegant animal (figure 3.12), fully meriting Walcott's choice for its specific name—*Marrella splendens.* Specimens measure from 2.5 to 19 mm (less than an inch) in length. The head shield is narrow, with two prominent pairs of spines directed backward (figures 3.13 and 3.14). Behind the head, twenty-four to twenty-six body segments, each bear a pair of biramous (two-branched) appendages (figure 3.15), composed of a lower walking leg and an upper branch bearing long and delicate gills (the source of Walcott's informal name, "lace crab"). A tiny button,

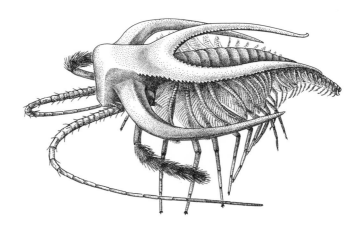

3.12. Side view of *Marrella*. Drawn by Marianne Collins.

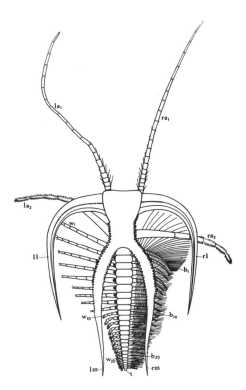

3.13. Reconstruction of *Marrella* by Whittington (1971), top view. Note the two pairs of appendages and the two pairs of spines on the head shield. The second pair of spines sweeps back to cover the entire organism. The gill branches are omitted on the animal's left side, and the leg branches on the right side—all for greater ease in visual resolution. These omissions are standard in scientific illustrations, but can be confusing if you don't know the tradition.

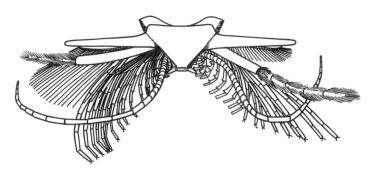

3.14. Front view of *Marrella*, seen as if walking right toward the reader (Whittington, 1971).

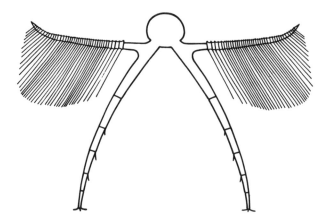

3.15. A pair of biramous appendages from *Marrella:* right and left gill branches above, leg branches below (Whittington, 1971).

called a telson, caps the rear end. Traces of the gut are preserved on some specimens. The rock surface just adjacent to the fossil itself often shows a characteristic dark stain—probably a remnant of body contents that oozed out beyond the external skeleton after death.

Harry worked for four and a half years on *Marrella,* personally preparing, dissecting, and drawing scores of specimens in varying orientations. Efforts of this sort are often left to assistants, but Whittington knew that he must do this basic work himself, over and over again, if he hoped to win a proper "feel" for Burgess preservation and its problems. This labor, however tedious and repetitious at times, also provided more than enough excitement to inspire perseverance. Harry spoke to me about his decision to do all the work himself, a commitment of several precious years in research:

> I think that it was vital. Of course it took hours and hours, but you saw everything yourself, and various things could sink in gradually. I love preparing [paleontological jargon for cleaning and exposing specimens in rock]. It is so exciting to find those hidden things. It is an incomparable thrill to reveal a hidden structure in the rock.

The Burgess studies of Whittington and his team are, for the most part, revisions, not first descriptions of newly found species. They are therefore presented in the context of previous interpretations and stand as evaluations of past work. Walcott had called *Marrella* a trilobite, or at least close enough to share the anatomical signature of the group. Størmer had made

Marrella a flagship of his Trilobitoidea, the sister-group of trilobites in his larger class Trilobitomorpha. Hence Whittington studied *Marrella* in the primary context of its relationship with trilobites, the subject of his lifelong expertise.

Whittington affirmed that the general form of *Marrella*'s body bears little overall resemblance to trilobites. The single head shield with its two prominent pairs of spines, the subsequent body, with so many uniform segments of gradually decreasing size, and the tiny button of a rear end—all scarcely recalled the "standard" trilobite, with an external skeleton usually shaped as a broad oval, and divided into three basic sections of cephalon, thorax, and pygidium (head, body, and tail for those who shun jargon).

But then, no one had ever invoked overall shape to make claims for *Marrella*'s affinity with trilobites. Størmer had cited a strong similarity in the biramous appendages of the body as a rationale for establishing his concept of Trilobitoidea. However, as Whittington studied hundreds of specimens, he slowly began to discover consistent, and probably fundamental, differences between the appendages of *Marrella* and those of all known trilobites. Whittington admitted, of course, that the basic structures are similar. This overall resemblance had never been doubted, and Whittington quoted Størmer's own words to emphasize the point: "These appendages are 'more or less trilobite-like' (Størmer, 1959, p. 26) in the general sense that there is a segmented walking leg and a filament-bearing gill branch" (Whittington, 1971, p. 21). But the differences began to impress Whittington even more. The walking leg of *Marrella*, with its six sections and terminal spines (see figure 3.15), bears one or two fewer segments than the standard and scarcely varying number in trilobites. Whittington concluded: "Neither branch is like that of any known trilobite, the walking leg having one (or two?) segments less than known in trilobites, the filament-bearing branch being differently constructed" (1971, p. 7).

Walcott's interpretation of the head shield and its appendages (1912 and 1931) had provided the strongest case for classifying *Marrella* as a trilobite. Trilobites (see inset, page 106) bear a characteristic, almost stereotypical, arrangement of appendages on the cephalon, or head shield—one pair (called antennae) in front of the mouth, and three pairs behind the mouth (older studies argue for four post-oral segments, but later work, especially Whittington's 1975 monograph on Burgess trilobites, has suggested three as more probable). Walcott reconstructed the head of *Marrella* in perfect conformity with the trilobite plan—one pair of antennae, and three subsequent pairs, which he called mandibles, maxillulae, and maxillae (1931, p. 31). Walcott even published photos (1931, plate 22) purporting to show this arrangement in clear and complex detail. This reconstruction provided a strong reason for linking *Marrella* with trilobites.

But Whittington soon developed doubts that gradually grew into disproof as he studied several hundred specimens. Later authors had not accepted Walcott's version. (Størmer, for example, who affirmed the link of *Marrella* with trilobites, rejected Walcott's reconstruction of the head, and relied on similarities in the body appendages.) Whittington found, first of all, that Walcott's illustrations were products of the retoucher's art, not fair maps of structures in rocks. On page 13, Whittington explains why his drawings of Walcott's specimens look so different from Walcott's 1931 photos: "The originals show that his illustrations were considerably retouched." By page 20, this measured assessment had yielded to one of the few acerbic remarks in all of Whittington's writings: "Several are heavily retouched to the point of falsification of certain features, notably the representation of the supposed mandible, maxilla, and maxillula."

Whittington found only two pairs of appendages, *both* pre-oral—in front of the mouth—attached to the head shield of *Marrella:* the long, many-jointed first antennae (equivalent to Walcott's "antenna" and interpreted by all in the same way), and a shorter and stouter pair of second antennae (Walcott's "mandible"), composed of six segments, several covered with setae, or hairs. Whittington could find no trace of Walcott's maxilla or maxillula, and he concluded that Walcott had confused some crushed and disarticulated legs of the first body segments with structures of the head shield. Walcott himself had admitted that he couldn't find these supposed appendages on most specimens: "The maxillulae and maxillae were so slender that they are usually absent as the result of having been torn off or crushed between the strong mandibles [Whittington's second antennae] and the thoracic limbs" (Walcott, 1931, pp. 31–32).

But recognition of two pre-oral (first and second antennae) and no post-oral appendages on the head shield of *Marrella* does not fully answer the anatomical question—for these two appendages could be related in a variety of potential ways, and a decision about taxonomic affinity depends upon the resolution. Whittington faced three major alternatives, all proposed before and each with different implications. First, the two antennae might represent the outer and inner branches of only one ancestral appendage—with the first antenna evolved from the outer gill branch (filaments lost and delicate shaft of numerous segments preserved), and the stout second antenna from the inner leg branch. Second, the two antennae might be truly separate by ancestry, arising as evolutionary modifications of two pairs of limbs on two original segments. Third, the second antenna, which looks so much like a walking leg, might really belong to the first body segment behind the head, and not be attached to the head shield at all. In this case, the head would bear only one pair of appendages—the first antennae.

Whittington wrestled with this issue above all others in resolving the anatomy of *Marrella*. He faced a technical problem because few, if any, specimens reveal the crucial point of connection between the head appendages and the shield itself. (The end of the appendage opposite to the point of attachment with the body—the distal, or farthest, end in technical parlance—is usually well preserved and easily visible because it projects well beyond the central axis of the body. But the end that attaches to the body—called the proximal, or nearest, end—is rarely resolvable because it lies under the axis and becomes inextricably mixed with the jumble of anatomical parts in this central region of the body.)

Whittington had to use all his tricks of analysis to resolve this issue— dissecting through the head shield to search for the limb attachments below, and seeking odd orientations that might reveal the proximal ends of the appendages. Figure 3.16 is a camera lucida sketch of the key specimen that finally drew Whittington to the second interpretation—the two antennae are distinct appendages, both attached to the head shield. This is

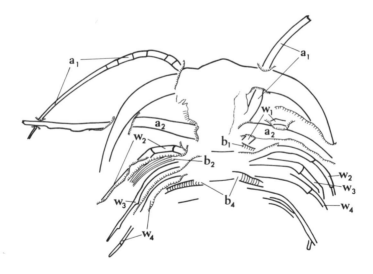

3.16. Camera lucida drawing of the key specimen of *Marrella* that settled the major problem in reconstructing the head anatomy. Only this specimen shows the two pairs of appendages (labeled a_1 and a_2) attached separately to the head shield.

the only specimen that clearly shows the proximal ends of both antennae, separately attached to the underside of the head shield.

Consider now the dilemma that Whittington faced as he began to compose his monograph on *Marrella*. He took for granted the old view that fossils fall within major groups and that life's history moves toward increasing complexity and differentiation. Yet *Marrella* seemed to belong nowhere. Whittington had found that the legs of the body segments are not sufficiently trilobite-like to warrant classification in this group. He had established a sequence of head appendages—two pre-oral and none post-oral—not only unlike the one pre-oral and three post-oral of trilobites but also completely unknown among arthropods. What was he going to do with *Marrella*?

Today, this situation would cause no problem. Harry would simply smile and say to himself—ah, another arthropod beyond the range of modern groups, another sign that disparity reached its peak at the outset and that life's subsequent history has been a tale of decimation, not increasing variety in design. But this interpretation was not available in 1971. The conceptual cart could not push this lead horse; in fact, the cart hadn't even been constructed yet.

In 1971, Harry was still trapped in the concept that Burgess fossils, as old, must be primitive—either generalized members of large groups that later developed more specialized forms, or even more distant precursors that combined features of several groups and could be interpreted as ancestors to all. He therefore toyed with the idea that *Marrella* might be a kind of precursor for both trilobites and crustaceans—trilobites for the vague similarity in leg structure, crustaceans for the characteristic two pairs of pre-oral appendages on the head shield. (A weak argument even in its own terms, for Whittington had shown important differences in detail between the legs of *Marrella* and those of trilobites, while crustaceans also have three post-oral appendages on the head shield, and *Marrella* has none.) Still, Whittington was stuck with a conventional notion of primitivity, and he could offer no more to *Marrella*. He wrote: "*Marrella* is one of the fossils indicating the existence of an early arthropod fauna, characterized by serially uniform, generally trilobite-like limbs . . . and by a lack of jaws, features associated with particle and detritus feeding" (1971, p. 21).

But Whittington still had to classify *Marrella*. Again, a quandary, for *Marrella* possesses unique features that violate the key characters of every group of arthropods. Harry, on the brink of a transforming insight, chose caution and tradition this one time—and placed *Marrella* in Størmer's Trilobitoidea, as the title of his monograph proclaims. Yet, as he did so, he felt the pain of betraying his own better judgment. "I had to put something at the top," he told me, "so I put 'Trilobitoidea.' " Yet, in the interval between submitting his manuscript and receiving printed copies, Whittington realized that he would have to abandon Trilobitoidea as an artificial group, a "wastebasket" hiding the most interesting story of arthropod evolution. He said to me: "When I saw *Marrella* printed with 'Trilobitoidea' on top, I knew it was a bust." But *Marrella*, in fact, had been the beginning of a boom—and the documentation of this anatomical explosion would soon transform our view of life.

Yohoia: A SUSPICION GROWS

On his cautious journey through the Burgess arthropods, Whittington meant to proceed in order of abundance. *Canadaspis* stood next in line, but Harry wanted a research student to handle the entire group of arthropods with bivalved carapaces (Derek Briggs would do this work with brilliant results, as Act 3 will show). *Burgessia* and *Waptia*, the two genera that Størmer had united as his subclass Pseudonotostraca, followed in terms of abundance. But Whittington had allocated these genera to his colleague Chris Hughes (who published a study of *Burgessia* in 1975, but has yet to finish his work on *Waptia*). Hence, Whittington tackled the next most abundant arthropod (with some four hundred specimens)—the interesting genus *Yohoia*, namesake of the national park that houses the Burgess Shale.

Whittington's second monograph, his 1974 study of *Yohoia*, marks a subtle but interesting transition in his thinking, a necessary step toward the major transformation to come. Whittington had struggled with *Marrella*, and had come to the correct empirical conclusion—that this most common Burgess genus fits into no known group of arthropods. But he lacked the conceptual framework for thinking of Burgess organisms as anything other than primitive or ancestral—and he certainly had no inclination to construct a new guidepost for only one example that might not be typical. But one is an oddity, and two a potential generality. With *Yohoia*, Whittington made his first explicit move toward a new view of life.

Yohoia is a very peculiar animal. It looks "primitive" and uncomplicated at first glance (figure 3.17)—an elongate body with a simple head shield, and no funny spines or excrescences. Walcott had placed *Yohoia* among the branchiopods, Størmer as an uncertain genus tacked to the end of Trilobitoidea. Yet, as Whittington proceeded, he became more and more puzzled. Nothing about *Yohoia* fitted with any known group.

The preservation of *Yohoia* left much to be desired by Burgess standards, and Whittington had difficulty resolving the order and arrangement

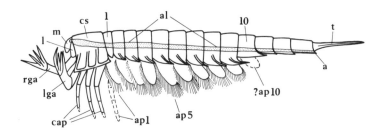

3.17. Reconstruction of *Yohoia* by Whittington (1974). Note the unique great appendage (labeled *rga* and *lga*) attached to the head.

of appendages—a crucial factor in arthropod taxonomy. He finally decided that the head probably bears three pairs of uniramous walking legs—nothing unconventional here, since this is the standard trilobite pattern and consistent with Størmer's placement in Trilobitoidea. But the most curious anomaly of all stands just in front—the large pair of grasping appendages, composed of two stout segments at the base and four spines at the tip. This design is unique among arthropods, and Whittington found no name in the panoply of available jargon. With elegant simplicity, he opted for the vernacular and called this structure the "great appendage."*

Yohoia bears no other appendages on its head shield—no antennae,† no feeding structures (the so-called jaws and mouth parts of insects and other arthropods are modified legs—the main source for our feelings of bizarreness or discomfort when we view films of enlarged insects eating). The first ten body segments behind the head bear lobate appendages fringed with setae, or hairlike extensions (figure 3.18; see also figure 3.17). The appendage on the first segment may have been biramous, including a walking leg as well—but Whittington was not able to resolve the appendages satisfactorily due to poor preservation. Segments 11–13 are cylindrical and carry

*Walcott, of course, had not failed to note this prominent organ, and its uniqueness did pose a problem for his conclusion that *Yohoia* was a branchiopod. Walcott evaded this dilemma by arguing that the great appendage was a male "clasper," or structure used to hold females during mating (and present in many branchiopods). But Whittington determined that all specimens bore great appendages, and disproved Walcott's rationale.

†Walcott had placed two species in the genus *Yohoia*—*Y. tenuis* and *Y. plena*. Whittington realized that the two animals are distinct and belong in different genera. *Y. plena,* which has antennae, is a phyllocarid, one of the arthropods with a bivalved carapace soon to be studied by Derek Briggs. Whittington removed this species from *Yohoia* and established a new genus, *Plenocaris*. *Yohoia tenuis* is the oddball, and subject of the 1974 monograph.

no appendages, while the last, or fourteenth, segment forms a flattened telson, or tail. Again, this arrangement of segments and appendages departs strongly from the standard trilobite pattern of biramous limbs on each body segment. *Yohoia,* with its great appendage in front, and curious arrangement of limbs behind, was an orphan among arthropods.

Whittington (interview of April 8, 1988) remembers his study of *Yohoia* as a turning point in his thinking. He had assimilated *Marrella,* despite its uniquenesses, under the two reigning *p*'s—"primitive" and "precursor." But *Yohoia* forced a different insight. This basically simple, elongate animal with many segments did have a primitive look in some respects. "This animal," he wrote, "resembles Snodgrass' hypothetical primitive arthropod in that the alimentary canal extended the length of the body" (1974, p. 1). But Whittington did not shunt the uniquenesses aside, particularly the form of the great appendage. He had attempted a reconstruction of *Yohoia* as a working animal—showing how the lobate body appendages with their setal fringes might have been used for swimming, for breathing (as gills), and for transporting food particles; and how the great appendage might have captured prey with its spiny tips and then folded back to bring food right to the mouth.

All these features were unique anatomical specializations that probably helped *Yohoia* to work in its own well-adapted way. This animal was not a precursor with a few oddities, but an entity unto itself with a mixture of primitive and derived characters. "In the exoskeleton and appendages," Whittington wrote, *"Yohoia tenuis* is clearly specialized" (1974, p. 1).

Thus, as the crucial year 1975 dawned, Whittington had completed

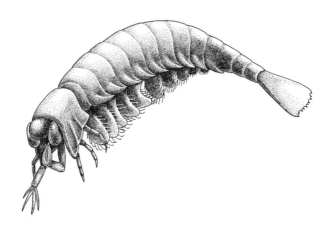

3.18. *Yohoia.* Drawn by Marianne Collins.

SYSTEMATIC DESCRIPTIONS

Class TRILOBITOIDEA Størmer, 1959?
Family YOHOIIDAE Henriksen, 1928
Genus *Yohoia* Walcott, 1912

3.19. The fateful first expression of doubt. Whittington (1974, p. 4) still placed *Yohoia* in the Trilobitoidea, but expressed his doubt about the status of Størmer's group.

monographs on two Burgess arthropods with the same curious result. *Marrella* and *Yohoia* didn't fit anywhere—and they were specialized animals apparently living well with their unique features, not simple and generalized creatures from the dawn of time, ripe for replacement by more complex and competent descendants.

Whittington remained too cautious to translate these suspicions into hard taxonomy. He still, and for one last time, placed *Yohoia* in Trilobitoidea, but with two crucial differences. He did not use Størmer's category in the title of his monograph, and he inserted a fateful question mark after the designation in his formal taxonomic chart (1974, p. 4)—the first overt sign of challenge to the old order (figure 3.19). Whittington wrote: "I am doubtful whether *Yohoia* should be placed in Trilobitoidea" (1974, p. 2). Never doubt the conceptual power of a question mark.

ACT 2. A New View Takes Hold:
Homage to *Opabinia*, 1975

Harry Whittington began his 1975 monograph on *Opabinia* with a statement that should go down as one of the most remarkable in the history of science: "When an earlier version of figure 82 [reproduced here as figure 3.20] was shown at a meeting of the Palaeontological Association in Oxford, it was greeted with loud laughter, presumably a tribute to the strangeness of this animal" (1975a, p. 1). Are you baffled by my claim? What is so unusual about this inoffensive sentence that doesn't even abandon the

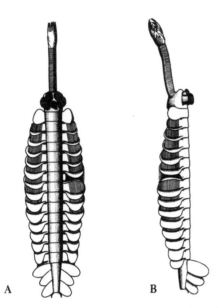

3.20.   Reconstruction of *Opabinia* by Whittington (1975). (A) Top view, showing the five eyes on the dorsal surface of the head. (B) Side view: note the orientation of the tail fins relative to the body; the dorsal surface is at the right.

traditional passive voice of scientific prose? Well, you have to know Harry Whittington, and you have to be steeped in the traditions of style for technical monographs. Harry, as I have stated many times, is a conservative man.* I doubt that he had, in all the several thousand pages of his output, ever written a personal statement, much less an anecdote about a transient event. (Even here, he could bring himself to do so only in the passive voice.) What, then, could possibly have persuaded Harry Whittington to begin a technical monograph in the *Philosophical Transactions of the Royal Society, London* with a personal yarn that seems about as fitting in this format as Kareem Abdul-Jabbar in Lilliput? Something really unusual was about to happen.

In 1912, Walcott had described *Opabinia* as yet another branchiopod

---

*I view this as a crucial and favorable feature for the general story of this book—because you can be sure that Whittington came to his new interpretation of the Burgess from an accumulating weight of evidence, not from any *a priori* desire to go down in history as a radical reformer.

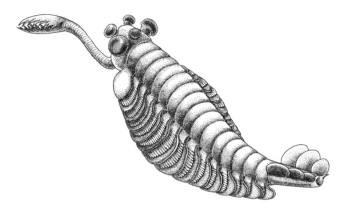

3.21. *Opabinia,* showing the frontal nozzle with terminal claw, five eyes on the head, body sections with gills on top, and the tail piece in three segments. Drawn by Marianne Collins.

crustacean. Its curious design, particularly the bizarre frontal nozzle (figure 3.21), had made *Opabinia* a center of Burgess attention. Many different reconstructions had been attempted, but all authors had found a place for *Opabinia* within a major group of arthropods. *Opabinia,* as the most puzzling of all Burgess arthropods, stood as a challenge and a logical next step for Harry Whittington after two monographs on common genera *(Marrella* and *Yohoia),* and one on the structure of trilobite limbs (1975b).

Whittington began his study of *Opabinia* without any doubt about its status as an arthropod. He soon received the surprise of his life, though the lesser oddities of *Marrella* and *Yohoia* had prepared him for astonishment from the Burgess. Whittington presented his first reconstruction of *Opabinia* to the annual meeting of the Palaeontological Association* in Oxford in 1972.

Laughter is the most ambiguous of human expressions, for it can embody two contradictory meanings. Harry recognized the laughter of his colleagues at Oxford as the sound of puzzlement, not derision—but it really shook him up nonetheless. Both Simon Conway Morris and Derek Briggs, his two superb students, agree that this Oxford reaction marked a turning point in Harry's work on the Burgess Shale. He simply had to resolve and diffuse that unanticipated and incongruous laughter. He had to overwhelm his colleagues with a reconstruction of *Opabinia* so incontro-

---

*The leading British professional association of paleontologists. They call themselves the "pale ass" for informal fun—a name even more humorous to an American, since the title only refers to a donkey in England (where your nether end is your arse).

vertible that all its peculiarities could pass into the realm of simple fact, and never again disturb the courts of science with the spirit of Milton's *L'Allegro:*

> Haste thee nymph, and bring with thee
> Jest and youthful Jollity, . . .
> Sport that wrinkled Care derides,
> And Laughter, holding both his sides.

Although *Opabinia* is a rare animal with only ten good specimens (Walcott found nine, and the Geological Survey of Canada added another in the 1960s), Walcott established its importance as a centerpiece in interpreting the Burgess fauna. He awarded *Opabinia* pride of place, describing this genus first among the Burgess arthropods (see table 3.1). Walcott put *Opabinia* at the head of his classification because he regarded the elongate body, composed of many segments without prominent and complex appendages, as "very suggestive of an annelidan ancestor" (1912, p. 163). Since the Annelida, or segmented worms (including terrestrial earthworms and marine polychaetes), are the presumed sister-group of the Arthropoda, an animal that combined characters of the two phyla might stand close to the ancestry of both and act as a link between these great invertebrate groups. To Walcott, *Opabinia* was the most primitive Burgess arthropod, the closest model for a true ancestor of all later groups.

But what arthropod features did Walcott discern in *Opabinia?* He had little to offer for the head, since he could find no appendages. The frontal "nozzle" might be interpreted as a pair of fused antennae, and the eyes were consistent with arthropod design (Walcott noted only two eyes, but Whittington found five—two paired and one central). Walcott admitted that "none of the heads . . . show traces of antennules, antennae, mandibles or maxillae. If these appendages were large they have been broken off; if small they may be concealed beneath the crushed and flattened large posterior section of the head" (1912, p. 168). I regard this statement as a lovely example of apparently unconscious bias in science. Walcott "knew" that *Opabinia* was an arthropod, so the animal had to have appendages on its head. Since he didn't find any, he provided explanations for their absence—either they were so large that they always broke off, or they were so small that they became hidden beneath the head. He never even mentioned the obvious third alternative—that you don't see them because they didn't exist.

(Walcott, by the way, also made another error—see the next paragraph—that may seem merely amusing or tangential but underscores the serious point that we observe according to preset categories, and often cannot "see" what stares us in the face. A set of empirical anomalies may have instigated the Burgess revision by Whittington and colleagues, but as we shall see, the conceptual framework of the new view, coalescing between 1975 and 1978, established a novel context that allowed further observations to be made. I preach no relativism; the Burgess animals are what they are. But conceptual blinders can preclude observation, while more accurate generalities guarantee no proper resolution of specific anatomies, but can certainly guide perceptions along fruitful paths.)

Walcott, following our primal biases of gender, found two specimens that appeared to lack the frontal nozzle. (Walcott thought that the nozzles were truly absent on these specimens, but Whittington later proved, by dissecting one of the specimens and finding the jagged edge of the break point, that the nozzles had been broken off.) On one specimen, Walcott found a slender, two-pronged structure in the same location as the nozzle. (This turned out to be a fragment from an unrelated worm, but Walcott interpreted it as a genuine part of *Opabinia*, in the same position as the nozzle of other specimens.) Walcott therefore concluded that he had discovered sexual dimorphism in *Opabinia*: the strong and stout nozzle belonging to the male (naturally), and the slender structure to the more delicate female. He wrote that these supposed females "differ from the male . . . in having a slender, bifid frontal appendage instead of the strong appendage of the male." He even foisted the stereotypes of active and passive upon his fictitious distinctions, arguing that the nozzle "was probably used by the male to seize the female" (1912, p. 169).

Walcott's main justification for regarding *Opabinia* as an arthropod lay in his interpretation of the paired body segments. He read these flaps as the gill branches of ancestrally biramous appendages. He thought that he had observed two or three "rather strong, short joints" (1912, p. 168) at the base of each flap, followed by the broad lobe bearing the gills. He hoped to find the inner leg branches as well, but he could never fully persuade himself, and eventually concluded that the walking legs probably existed in an "insignificant or rudimentary" form (1912, p. 163).

Walcott was clearly troubled by the failure of *Opabinia* to preserve any smoking gun of arthropod affinity. He even took some modern anostracans and crushed them between plates of glass, trying to simulate the conditions of Burgess fossilization. This mayhem provided some solace, because such treatment often destroyed all evidence of the delicate appendages. He wrote: "After flattening specimens of *Brachinecta* and *Branchipus* between plates of glass and studying them, I am greatly surprised that any distinct characters of the appendages are preserved in the fossils in a recognizable condition" (1912, p. 169). Walcott had shown the cardinal skill of his adopted profession—administration. He had put the best face upon adversity. *Opabinia* would remain an arthropod.

But Walcott had been downright circumspect compared with later reconstructions that added more and more arthropod features with less and less compunction. In 1931, the great ecologist G. Evelyn Hutchinson, driven to paleontology by the fascinating problem of how anostracans could change their environmental preferences from Cambrian oceans to modern freshwater ponds, reconstructed *Opabinia* in the standard upside-down position of a swimming anostracan (figure 3.22). He turned the lateral flaps into long bladelike appendages neatly fitted to the side of an arthropod carapace.

The climax of this imaginative tradition arrived with the aesthetically lovely but fanciful reconstruction of Simonetta (1970).* *Opabinia* has become an ideal arthropod (figure 3.23). The frontal nozzle is shown with a longitudinal suture (entirely imaginary), indicating its origin as a pair of antennae, now fused. Simonetta "found" two additional pairs of short arthropod appendages on the head—one constructed from a pair of eyes, the other from a bump on the carapace. On each segment of the body itself, Simonetta drew a strong and fully biramous appendage—a bladelike gill branch above a small but firm leg branch. Whittington faced this unchallenged tradition when he began his work on the ten precious specimens of *Opabinia*.

I now come to the fulcrum of this book. I have half a mind to switch to upper case, or to some snazzy font, or to red type, for the next page or two—but I desist out of respect for the aesthetic traditions of bookmaking. I also refrain because I do not wish to fall into the lap of legend (having already dispersed one for the discovery of the Burgess Shale). My emotions and desires are mixed. I am about to describe the key moment in this drama, but I am also committed to the historical principle that such moments do not exist, at least not as our legends proclaim.

Key moments are kid stuff. How can such a story as this, involving so

*A. M. Simonetta, an Italian paleontologist, deserves a great deal more credit than this book has space to provide. He alone, after Walcott and before Whittington, attempted a comprehensive program of revision for Burgess arthropods. He worked as Walcott had, and with Walcott's specimens, treating the fossils essentially as films on the rock surface and attempting no preparation of specimens. He therefore made many mistakes in a long series of papers published during the 1960s and 1970s. But he also provided substantial improvements upon several earlier studies, and through his comprehensive efforts reminded paleontologists about the richness of the Burgess Shale. Since science is a process of correction, Simonetta's errors also provided an important spur to Whittington and his colleagues.

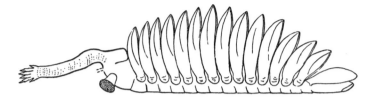

3.22. Hutchinson's reconstruction of *Opabinia* as an anostracan swimming upside down in the modern position (1931).

many people engaged in complex intellectual struggles, proclaim any moment as a single focus, or even as most important? I have labored to master all the details and to arrange them in proper order. How can I now blow all this effort on the myth of eureka? I suppose that one can discover a single object—say, the Hope diamond—at a particular moment, but even such a pristine event has a tangle of inevitable antecedents in geological training, political intrigue, personal relations, and good luck. But I am talking about an abstract and far-reaching transformation in our view of life's pattern and the meaning of history. How can such a complex change possess a moment before, when it wasn't, and a moment after, when it was? Does natural selection, or laissez-faire economics, or structuralism, or the rationale for the Immaculate Conception of Mary, or any other complex moral or intellectual position, owe its formulation to a single person, place, or day?*

Still, as Orwell said about his metaphorical Russia in a farmyard, some animals are more equal than others. We need heroic items and moments to focus our attention—the apple that hit Newton and the objects that Galileo did not drop from the Leaning Tower. The beat goes on, but we may discern a high spot in the continuity.

I believe that the transformation of the Burgess Shale did have a Rubicon of sorts, at least symbolically—a key discovery that can separate a before and an after.

---

*My Catholic friends may cite Pius IX and December 8, 1854, for the last item in my list, but *Ineffabilis Deus* was an official resolution under the rules of the institution, and no one could pick one moment as paramount in a millennium of previous debate. On Darwin's long and complex struggle to develop the theory of natural selection, see Howard Gruber, *Darwin on Man.* (New York: Dutton, 1974).

So we return to Harry Whittington, facing the entire world's supply of *Opabinia*. Everyone had always identified this animal as an arthropod, but no one had found the smoking gun, the segmented appendages that define the group. But then, no one before Whittington had possessed the techniques needed to seek out small appendages hidden under an external carapace. A few years before, Harry had made the central methodological discovery that the Burgess Shale fossils are three-dimensional objects (however crushed), with top layers that one can dissect away, to reveal the structures underneath. Harry had already resolved *Marrella, Yohoia,* and the Burgess trilobites with this method.

Opabinia virtually clamored for its crucial experiment under the new techniques: dissect through the carapace to find the body appendages and their attachments, dissect through the head shield to find the frontal appendages. So Harry dissected, in full confidence that he would find the jointed appendages of an arthropod. Harry dissected—*and he found nothing under the carapace.*

Opabinia was not an arthropod. And it sure as hell wasn't anything else that anyone could specify either. On close inspection, nothing from the Burgess Shale seemed to fit into any modern group. *Marrella* and *Yohoia* at least were arthropods, even if orphaned within this giant phylum. But what was *Opabinia*?

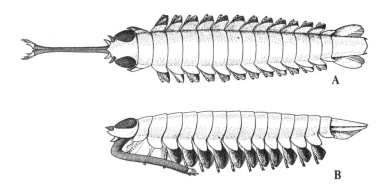

3.23. Attractive but fallacious restoration of *Opabinia* as an arthropod by Simonetta (1970). (A) Top view. (B) Side view. Simonetta showed the frontal nozzle as formed by fused antennae, and drew biramous appendages on each supposed body segment.

Whittington's conclusion may have been confusing, but it was also liberating. *Opabinia* did not have to conform to the demands of arthropod, or any other, design. Whittington could come as close as any paleontologist ever had to the unattainable ideal of Parsifal—the perfect fool, with no preconceptions. He could simply describe what he saw, however strange.

Opabinia is peculiar indeed, but not inscrutable. It works like most animals. *Opabinia* is bilaterally symmetrical. It has a head and a tail, eyes, and a gut running from front to back. It is an ideal creature for any eager scientist—not so crazy as to be intractable, but weird enough to thrill any curious person.

Whittington began his monograph by chiding his predecessors for their unquestioning allegiance to the arthropod model, and for their consequent tendency to rely more on expectations of the model than on observation of the specimens: "Continuous interest in *Opabinia* has not been accompanied by critical study of the specimens, so that fancy has not been inhibited by facts. The present work aims to provide a sounder basis upon which to speculate" (1975a, p. 3). With characteristic understatement (his personal tendency added to the British norm), Whittington then wrote: "My conclusions on morphology have led to a reconstruction which differs in many important respects from all earlier ones" (1975a, p. 3).

These "many important respects" led to an animal that might grace the set of a science-fiction film, if considerably enlarged beyond its actual length of 43–70 mm (less than three inches at most). Consider the major features of Whittington's reconstruction:

1. *Opabinia* does not have two eyes, but, count 'em, five! These are arranged as two pairs on short stalks, with a fifth eye, probably unstalked, mounted on the mid-line (see figure 3.20).

2. The frontal nozzle is not a retractable proboscis or a product of fused antennae (the two favorite interpretations consistent with arthropod design). It is attached to the bottom front border of the head and extends forward. It is a flexible organ, built as a cylindrical striated tube—literally like the hose of a vacuum cleaner, and perhaps bendable by the same principles. Its end is divided longitudinally into two halves, each with a group of long spines directed inward and forward. The tube may have contained a central, fluid-filled canal—a good device for requisite stiffness with enough flexibility.

3. The gut is a single tube running straight along the center of the animal for most of the body's length (see figure 3.24). However, at the head, the gut makes a U-shaped bend, and turns sharply around to produce a backward-facing mouth. Interestingly, the frontal nozzle has just the right length to reach, and appropriate flexibility to bend around and pass

food to, the mouth. Whittington suggests that *Opabinia* fed primarily by capturing food in the "pincers" formed by the spiny parts at the front of the nozzle, and then bending the nozzle around to the mouth.

4. The main portion of the trunk has fifteen segments, each segment bearing a pair of thin lateral lobes, one on each side of the central axis. These lobes overlap, and are directed downward and outward (see figure 3.20).

5. Each lobe except the first bears on its dorsal surface a paddle-shaped gill attached near the base of the lobe. Although the bottom surface of the gill is flat, the upper surface consists of a set of thin lamellae, overlapping like a deck of cards spread out.

6. The last three segments of the trunk form a "tail" built by three pairs of thin, lobate blades directed upward and outward (see figure 3.20).

Whittington needed all his special methods of dissection, varied orientations, and part–counterpart to resolve the morphology of so peculiar a beast. He also discovered that a failure to appreciate these methods had provided a major argument to support the arthropod model. Walcott had confused part and counterpart in one important specimen. He thought that he was viewing the bottom surface of the animal; in fact, he was looking down upon the upper surface. Raymond, accepting this upside-down interpretation, had made the perfectly reasonable claim that the gills of *Opabinia* lay *below* the outer carapace—as in the standard arthropod arrangement, with gill branches as the upper limbs of biramous appendages located just under the carapace. But in the correct orientation, the gills lie above the body lobes in a most unarthropod-like orientation.

Figures 3.24–3.26 provide a striking illustration of the power of Whittington's methods. These are his camera lucida drawings of three specimens, in varying orientations, each combining features from the part and counterpart of the same specimen. Figure 3.24 provides a view from above (dorsal). We see the position of the eyes and nozzle, the full sequence of lateral lobes, and the gills lying above the lobes. The gut runs as a straight tube down the middle of the body. Figure 3.25 is a side view and reveals several features that could not be seen from the top. We now discern the point of attachment for the nozzle, and we note that the gut bends in a U to form the rearward-facing mouth. (In top view, the bend and rearward section collapse upon the straight portion and cannot be distinguished at all.) The top view also tells us nothing about the relative positions of lateral lobes and tail fins, for these are collapsed into the same plane. But the side view of figure 3.25 shows the lateral lobes pointing downward and away from the body, while the tail fins stand high and point upward—in good positions, respectively, for oars and rudders.

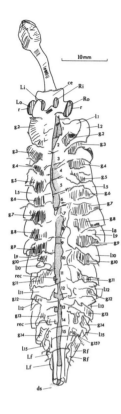

3.24. Camera lucida drawing for a specimen of *Opabinia* in the conventional position, viewed from the top. On each side, gills (labeled *g*) and lobes *(l)* are clearly distinguishable; the trace of the gut runs along the mid-line. Two pairs of eyes are visible, and the nozzle extends forward from the front end.

Figures 3.24 and 3.25 provide the two basic orientations, but they still leave several questions unanswered—and further specimens are needed. For example, neither shows the full complement of five eyes (they are delicate, and often collapse together into a jumble). Figure 3.26 fills some crucial gaps: five separate eyes are visible, and the frontal nozzle bends around to the area of the mouth.

Marrella and *Yohoia* had challenged Walcott's shoehorn, but these genera were only orphaned within the Arthropoda. With *Opabinia*, the game cranked up to another level, and changed unalterably and forever. *Opabinia* belonged nowhere among the known animals of this or any former earth. If Whittington had chosen to place it within a formal classification at all (he wisely declined), he would have been forced to erect a new phylum for this single genus. Five eyes, a frontal nozzle, and gills above lateral flaps! Walcott's shoehorn had fractured. Whittington wrote with characteristic brevity in the passive voice: "*Opabinia regalis* is not considered to have been a trilobitomorph arthropod, nor is it regarded as an annelid" (1975, p. 2). Harry may be a measured man, but he knew what

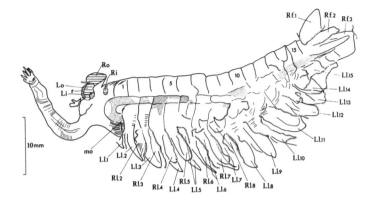

3.25. A specimen of *Opabinia* preserved in a more unusual orientation, on its side. Here lobes and gills of the right and left sides are jumbled together and difficult to distinguish. But many features not visible in the conventionally positioned specimen of figure 3.24 can now be understood: the orientation of the tail fins (labeled *Rf.1–Rf.3*) relative to the side lobes, the point of insertion for the nozzle, and the rearward bending of the front end of the gut.

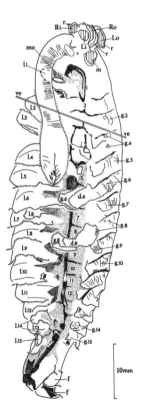

3.26. A third specimen of *Opabinia*, again in the conventional position. Several features not apparent in the other specimens can be distinguished: the fifth eye (labeled *m*, for "middle eye") is visible at the upper right, and we note that the nozzle can bend around to the level of the mouth.

Opabinia implied for the rest of the Burgess fauna. "The Burgess Shale," he remarked laconically, "contains other undescribed segmented animals of uncertain affinities" (1975, p. 41).

I believe that Whittington's reconstruction of *Opabinia* in 1975 will stand as one of the great documents in the history of human knowledge. How many other empirical studies have led directly on to a fundamentally revised view about the history of life? We are awestruck by *Tyrannosaurus;* we marvel at the feathers of *Archaeopteryx;* we revel in every scrap of fossil human bone from Africa. But none of these has taught us anywhere near so much about the nature of evolution as a little two-inch Cambrian odd-ball invertebrate named *Opabinia.*

ACT 3. The Revision Expands: The Success of a Research Team, 1975–1978

SETTING A STRATEGY FOR A GENERALIZATION

Think of all the accumulation songs in the English folk tradition. The first item never amounts to much—a partridge in a pear tree, or a paper of pins. "Green Grow the Rushes, Ho" puts it best: "One is one and all alone and ever more shall be so."

Opabinia carries the full weight of the Burgess message for a new view of life. It is as bizarre, as different from all living creatures, as anything else in the Burgess Shale. But one is all alone and ever more shall be so. The fossil record contains other oddities here and there—like the Tully Monster of Mazon Creek (see page 63). *Opabinia,* just one case, is a shrug of the shoulders, not a discovery about life in general. This example did not establish an incontrovertible new interpretation. Quite the opposite; it only hinted at a possibility worth exploring—especially with *Marrella* and *Yohoia* indicating that something similar, at a lower level, was running rampant among the Burgess arthropods.

All interesting issues in natural history are questions of relative frequency, not single examples. Everything happens once amidst the richness of nature. But when an unanticipated phenomenon occurs again and again—finally turning into an expectation—then theories are overturned. *Opabinia* would not earn its status as primer and flagship for a new view of life until its message of taxonomic uniqueness became ordinary within the Burgess Shale, however exquisitely rare for later times.

This need for numbers of examples—for an assessment of the relative frequency of oddballs within the entire Burgess fauna—makes the myth of the hero, grade B Western movie style, inapplicable to this story in principle. Harry Whittington could not be a lone lawman subduing saloonful after saloonful of reprobates. *Marrella* had taken more than four years. The Burgess arthropods alone would require several lifetimes. Whittington could either intone the lament of the frustrated Mercedes—"So many pedestrians, so little time"—or he could enlist a fleet to help. He chose the second alternative. Science is a collective enterprise in any case.

After selecting the genera that would provide a focus for his personal studies, Whittington divided the remaining arthropods into three groups, each suitable for an extensive research project by a collaborator. In addition, and growing both more troubling and crucial since the identification of *Opabinia* as an oddball outside any established phylum, stood the many genera that Walcott had classified as annelid worms (1911c). If Walcott's shoehorn had hidden a general theme of taxonomic uniqueness, the story would probably emerge (if not explode) even more clearly from the annelids than from the arthropods. Arthropods have clear and complex defining characters. Walcott might have wrongly shoehorned his arthropods into conventional groups within the phylum, but most were genuine arthropods at least (with *Opabinia* and, later, *Anomalocaris* as exceptions). But anything soft, segmented, and bilaterally symmetrical might be called a worm. The potential for oddballs loomed largest among Walcott's "annelids."

Whittington doubted that the three arthropod groups were coherent taxonomic assemblages. Each shared some features of superficially similar appearance, but *Marrella* and *Yohoia* had already taught caution about such externalities. Still, the three groups formed convenient divisions for research efforts, and the postulate of coherence could become a focal question for testing. (All three groups turned out to be heterogeneous—an important conclusion that confirmed the status of Burgess arthropods as spectacularly disparate compared with all later faunas.)

The three groups, all generally recognized in Burgess classifications from Walcott to Størmer, were (1) the large assemblage of arthropods with bivalved carapaces, always assumed to be true malacostracan crustaceans; (2) the "merostomoid" species, generally oval in shape and with a large discrete head shield that seemed to recall the great group of fossil eurypterids and their cousins the horseshoe crabs; and (3) apparent crustaceans with simple carapaces not divided into two parts, or valves.

When Whittington began his work in the late 1960s, two junior colleagues agreed to take on the smaller projects in this list. David Bruton of the University of Oslo received the "merostomoids" (I have discussed his work on *Sidneyia* in my section on techniques, early in chapter III, and shall report his conclusions in proper chronological sequence, in Act 5). Chris Hughes of Cambridge tackled *Burgessia* and *Waptia,* third and fourth most common Burgess arthropods, and forming the group of apparent crustaceans with simple carapaces. The monograph on *Waptia* has yet to appear, but Hughes's 1975 treatment of *Burgessia* provided an important affirmation of the growing pattern already indicated by *Marrella* and *Yohoia. Burgessia,* with its oval carapace, and long tail spike (almost twice the length of the body), was not a notostracan branchiopod, as Walcott had believed, but yet another arthropod orphan of unique design (figure 3.27). Hughes declined to make a formal taxonomic place for *Burgessia,* because he regarded this genus as a peculiar grabbag, combining features generally regarded as belonging to a number of separate arthropod groups. He concluded:

> Since the current restudy of all the Burgess Shale arthropods is revealing that the detailed morphology of these forms is not as previously thought, the present author considers further discussion of the affinities of *Burgessia* as premature. . . . What is apparent from this restudy is that *Burgessia* did

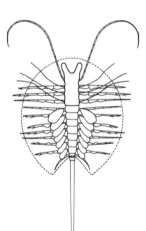

3.27. Reconstruction of *Burgessia* by Hughes (1975).

possess a mixture of characters . . . many of which are to be found in modern arthropods of various groups (1975, p. 434).

The arthropod story was becoming more and more curious.

MENTORS AND STUDENTS

Universities operate one of the few survivors of the old apprenticeship system in their programs for awarding doctoral degrees. Consider the anomaly. You spend your entire educational career, from kindergarten to college, becoming more and more independent of the power of individual teachers; (cross your first-grade teacher and your life can be hell for a year; displease a college professor, and the worst you can do is fail a single course). Then you become an adult, and you decide to continue for a Ph.D. So what do you do? You find a person whose research intrigues you, and sign on (if he will accept and support you) as a part of a team.

In some fields, particularly those with large and expensive laboratories dedicated to the solution of definite problems, you must abandon all thought of independence, and work upon an assigned topic for a dissertation (choice in research is a luxury of later postdoctoral appointments). In more genial and individualistic fields like paleontology, you are usually given fair latitude in choosing a topic, and may emerge with a project uniquely your own. But in any case you are an apprentice, and you are under your mentor's thumb—more securely than at any time since the early years of primary school. If you and he have a falling out, you quit, or pack up and go elsewhere. If you work well together, and your mentor's ties to the profession are secure, you will get your degree and, by virtue of his influence and your proven accomplishments, your first decent job.

It's a strange system with much to criticize, but it works in its own odd way. At some point, you just can't proceed any further with courses and books; you have to hang around someone who is doing research well. (And you need to be on hand, and ready to assimilate, all the time, every day; you can't just show up on Thursday afternoon at two for a lesson in separating parts from counterparts.) The system does produce its horrors—exploitive professors who divert the flow of youthful brilliance and enthusiasm into their own dry wells, and provide nothing in return. But when it works (as it does rather more often than a cynic might expect, given the lack of checks and balances), I cannot imagine a better training.

Many students don't understand the system. They apply to a school because it has a general reputation or resides in a city they like. Wrong, dead wrong. You apply to work with a particular person. As in the old apprenticeship system of the guilds, mentor and student are bound by mutual obligations; this is no one-way street. Mentors must, above all, find and provide financial support for students. (Intellectual guidance is, of course, more fundamental, but this part of the game is a pleasure. The real crunch is the search for funding. Many leading professors spend at least half their time raising grant support for students.) What do mentors get in return? This reciprocation is more subtle, and often not understood outside our guild. The answer, strange as this may sound, is fealty in the genealogical sense.

The work of graduate students is part of a mentor's reputation forever, because we trace intellectual lineages in this manner. I was Norman Newell's student, and everything that I ever do, as long as I live, will be read as his legacy (and, if I screw up, will redound to his detriment—though not so seriously, for we recognize a necessary asymmetry: errors are personal, successes part of the lineage). I happily accept this tradition and swear allegiance to it—and not for motives of abstract approbation but because, again as with the old apprenticeship system, I get my turn to profit in the next generation. As my greatest joy in twenty years at Harvard, I have been blessed with several truly brilliant students. The greatest benefit is an exciting lab atmosphere for the moment—but I am not insensible to the custom that their future successes shall be read, in however small a part, as mine also.

(By the way, this system is largely responsible for the sorry state of undergraduate teaching at many major research universities. A student belongs to the lineage of his graduate adviser, not to the teachers of his undergraduate courses. For researchers ever conscious of their reputations, there is no edge whatever in teaching undergraduate courses. You can do it only for love or responsibility. Your graduate students are your extensions; your undergraduate students are ciphers in your fame. I wish that this could change, but I don't even know what to suggest.)

This system is even more exaggerated in England. In the United States you apply through a department to work with an adviser. In England, you apply directly to a potential mentor, and he secures the funds, almost always earmarked for particular projects. Harry Whittington knew that the ultimate success of the Burgess project—its expansion from the detailed description of a few odd animals to an understanding of an entire fauna—depended upon graduate students. Of the two ingredients, he could influence one—the garnering of funds; for the other he could only pray to the goddess of good fortune—the interest of brilliant students.

Harry did his job on the first score. He had two projects outstanding (in both senses of that word)—bivalved arthropods and "worms." He secured funding for two students—for one, from government grants, and for the other, from private monies administered by his college, Sidney Sussex.

Lady Luck came through on the second score (with a boost from Harry's own successes, for good students keep their eyes open and gravitate toward mentors doing the most exciting work). In 1972, at exactly the right stage in the flow of Burgess developments, events disproved my cherished theory of academic spacing—that brilliant students come but once in five years (since five years is the usual length of graduate study, you never have more than one at a time for very long). At the same time, Harry Whittington—lucky, lucky man—received applications from two brilliant students: Derek Briggs, an Irishman who had done undergraduate work at Trinity College, Dublin; and Simon Conway Morris, a Londoner who had just completed his first degree at Bristol University (where Harry had sat, as external examiner, on the committee to judge his undergraduate thesis). From then on, however restricted the daily contact, and despite an in- dividuality in working styles that precluded any cohesive research group, the Burgess work became a joint effort of three increasingly equal part- ners—Briggs, Conway Morris, and Whittington (in nonjudgmental alpha- betical order), three men with a common purpose and a common set of methods, but as different as could be in age and in general approaches to science and life.

Harry Whittington knows the rules and the score. In our conversations, he has emphasized above all else, and with no false modesty, that the Burgess revision became a complete and coherent project—not just a se- quence of monographs—when he secured the dedication of Briggs and Conway Morris. For he could then forge a goal that he might live to complete, and not, like the architect of a medieval cathedral, just draft a blueprint and lay a foundation, but never hope to see the entire building.

CONWAY MORRIS'S FIELD SEASON IN WALCOTT'S CABINETS: A HINT
BECOMES A GENERALITY, AND THE TRANSFORMATION SOLIDIFIES

Odd couples are a staple of drama and comedy. Conservative intellectuals of quality will often embrace radical students with outlandish life styles because they sense the light of brilliance and nothing else then counts. Bernie Kummel, who threatened to take a rubber hose to radical students in the 1970s, and who despised (and feared) any eccentricity of manner or dress, loved Bob Bakker (then our student, now the spearhead of new ideas about dinosaurs) like a son, despite his shoulder-length hair and radical notions about absolutely everything. (Bernie's judgment was not always so good. At one time, he and Harry Whittington formed the invertebrate- paleontology group at Harvard. Bernie regarded Harry as too traditional, and was pleased when he chose to leave for Cambridge. Bernie then hired me as a very junior replacement. Not much of a trade.)

Simon Conway Morris, who described himself to me as "bloody-minded as a teen-ager, and usually antisocial," struck Whittington as the best candidate for the craziest of all Burgess challenges—Walcott's "worms." Simon's teachers at Bristol had described him to Harry as a man who "sits in the corner of the library reading, and wears a cloak." Harry remembers his first reaction to this news: "The anarchist, I thought . . . Oh Lord." But Harry had also sensed the spark of brilliance, and as I said, nothing else really counts.

Worms presented both the biggest headache and the greatest promise for a project now explicitly searching for oddballs since the resolution of *Opabinia*. For if oddballs existed in abundance, previous investigators would have shoveled most misfits into the old category Vermes, or "worms." Worms are the classic garbage-pail group of taxonomy—the slop bucket for the dribs and drabs (Simon calls them "odds and sods") that don't fit anywhere, but need to be shunted someplace when you are trying to landscape the estate into rigorous order. Worms have played this role ever since Linnaeus himself, who shoved a remarkably heterogeneous group of creatures into his Vermes. Most animals are basically elongate and bilaterally symmetrical. So if a creature displays this form, and you don't know what it is, call it a worm.

Harry, a remarkably kind man, trembled at the idea that he might be ending a promising career at the beginning by giving such an intractable project to a greenhorn. To this day, he seems almost wracked with anxiety when he remembers what he did—even though the results have been spectacular: He reminisced to me: "With fear and trepidation, I suggested this to Simon. . . . I felt awful; of all ghastly things to start a research student on! Gosh, how could I dare to do that to anybody? Yet I had a wild hunch he could do it."

Simon was delighted; he has been running ever since. The solid center-pieces of this project are his two fine monographs on Burgess worms that truly belong to modern phyla—the priapulids (1977d) and the polychaetes (1979). I shall discuss these works in their proper sequence. But Simon did not begin with this conventional material; would you really expect such a traditional start from a man who wears a cloak and won't come to morning coffee?

In the spring of 1973, Whittington sent both Briggs and Conway Morris to Washington to draw Walcott's "type" specimens (the ones used in the original descriptions of the species, and the official bearers of Walcott's names), and to select specimens for loan to Cambridge. An old saying, attributed to Pasteur, proclaims that fortune favors the prepared mind.

Simon, a man of ideas, had chosen to work with Harry, and reveled at receiving the worms as a project, because he sensed that the prospect for a larger message from the Burgess centered upon the documentation of oddballs—both their anatomy and their relative frequency. *Opabinia* had forced its attention upon Harry. Simon, in stark contrast, went hunting for Burgess oddballs. "I have a natural temptation to emphasize the unusual," Simon told me. "A new brachiopod from Northern Ireland is no competition for a new phylum."

Imagine the situation, and the opportunity. Simon faced some eighty thousand specimens in Walcott's collection. Most had never been described, or even gazed upon. No one had ever examined this treasure with the idea that taxonomic oddballs might abound. So Simon did something both simple and obvious in concept, yet profoundly different from any previous approach to the Burgess—and therefore courageous. He went on a protracted fishing expedition in the Smithsonian drawers of Burgess material. He opened every cabinet and looked at every slab, consciously searching for the rarest and most peculiar things he could find. The rewards were great, the success almost dizzying. At first, you jump up and down; after a while, the richness benumbs you. By the time he found *Odontogriphus* (see page 147), he could only say to himself, "Oh fuck, another new phylum."

I cannot imagine a greater contrast (and, therefore, better seeds of drama) than the disparate styles of Whittington and Conway Morris— Harry, the older conservative systematist, about to start the greatest project of a full life, versus Simon, the radical beginner, consciously seeking to overturn established opinion. Their working procedures could not have been more different. Harry began with greatest caution, choosing the most common animal in the Burgess. He proceeded with a series of monographs on individual genera, each taking years of preparation: *Marrella* (1971), *Yohoia* (1974), trilobite limbs (1975b), *Opabinia* (1975a), and as we shall see, *Naraoia* (1977) and *Aysheaia* (1978). He confined his work (or so he thought when he began) to the arthropods, the group that he knew best. He started with conventional views about the taxonomy of Burgess organisms, changing his mind only when unexpected evidence forced itself upon his consciousness. Simon, by contrast—with the innocence of Pearl Pureheart and the proven skill of Alvin Allthumbs, but armed with the sublime confidence of Muhammad Ali as his youthful avatar Cassius Clay—began with an explicit search for embodiments of the most radical interpretation of Burgess anatomy. The rarer the better; several of Simon's weird wonders are reconstructions based upon single specimens. In two years, 1976 and 1977, Conway Morris initiated his career by publishing five short papers, on five creatures with the anatomical uniqueness of new phyla.*

Such differences should breed dissension and open conflict. Nothing of the kind occurred—intellectual drama of the highest order, yes, but no juicy stories of overt battle. Oh, Derek does remember Harry mumbling a bit about people running before they learned to walk, and some private feelings may be left unsaid to this day. But when I asked Harry how he felt about a student who published five short papers before his Ph.D., sometimes basing new phyla upon single specimens, he replied: "I stood by and smiled. I wouldn't dream of discouraging a research student."

I know that the following comment is trite, but the foundation of banality is often evident truth: The final coalescence of the Burgess transformation emerged from a lovely synergism between these two disparate approaches. Perhaps the process of interpretation would have led to the final outcome in any case. Perhaps either the slow sequence of descriptive monographs or the rapid succession of short papers with radical claims would eventually have compelled assent by itself. But nothing can beat the one–two punch of laborious description so careful that it cannot be gainsaid combined with overt claims so sparsely documented and so divergent from tradition that they can only inspire fury—and attention. I know that this combination "just happened" along one of the odd and unpredictable pathways of human affairs, but if anyone is up there regulating the progress of knowledge, he could not have acted with better or more deliberate purpose than by arranging this synergism of youth and experience, caution and daring.

I stopped the narrative once before (with *Opabinia*) to announce a key moment meriting special type for emphasis, and I shall do so just once

---

*Since Simon and Derek began working with Harry Whittington in 1972, the year of the infamous laughter over *Opabinia* at the Oxford meeting, I had assumed that their prodding must have convinced Harry to take the drastic step of declaring *Opabinia* as a unique anatomy of phylum-level status. This is how the script is supposed to go—the Young Turks dragging the old farts into the light of exciting modernity. Terrible screenplay, not at all like complex life. Simon may be ideologically radical, but he is one hell of an excellent descriptive anatomist—and anyone who would be fooled enough by externalities to rank Harry as an old fart understands nothing about the multifarious nature of genius. In any case, all three protagonists assure me that Harry worked out the interpretation of *Opabinia* without any hectoring or encouragement from radicals on the sidelines. The converse is equally true and contrary to script. Harry neither discouraged Simon as he wrote his five papers, nor helped with frequent counseling. Harry played virtually no role in Simon's first forays. He can remember only one intervention—an insistence that Simon use his techniques of dissection to excavate the spines of *Hallucigenia* right to the point of their connection with the body. Damned good advice, but scarcely the stuff of general guidance.

again (for *Anomalocaris*); but Simon's field season in the cabinets of the Smithsonian marks the second of three major transitions, as I read the story of the Burgess. When Simon began, *Opabinia* was hinting at something strange, but no one knew either the extent or the nature of the phenomenon; I believe that Harry was still favoring an interpretation of oddballs as stem groups, primitive combinations of characters that would later sort themselves into discrete phyla living today, rather than as uniquely specialized experiments in multicellular design, separate lineages without later issue. When Simon completed his initial sequence of five papers on curiosities, the tentative and peculiar had become a Burgess norm, and the notion of separate lineages beyond the realm of modern anatomy had displaced the conventional fallback to "primitive" and "precursor." Whittington recalled his gradually dawning reaction to Simon's discoveries: "The whole atmosphere changed. We were not just dealing with predecessors of known groups. The whole thing was beginning to make a picture."

Simon's five oddballs span a remarkable range of anatomy and life style. Their only common theme is peculiarity.

1. *Nectocaris*. Walcott did single out this peculiar animal, represented by only one specimen lacking a counterpart—for Conway Morris found a photo, retouched as usual, next to the well-prepared specimen. But Walcott had published nothing, and left no notes. Conway Morris justified his decision to publish on such scant information: "The fine preservation and unusual anatomy warrant notice being taken of this unique specimen" (1976a, p. 705).

From the "neck" forward, *Nectocaris* looks mostly like an arthropod (figure 3.28). The head bears one or two pairs of short, forward-projecting, but apparently unjointed (and therefore not arthropod-like) appendages. A pair of large eyes, probably borne on stalks, lies just behind. The back part of the head is enclosed by a flattened oval shield, perhaps bivalved. But the rest of the body evokes no particular hint of arthropod, and gives off more than an intriguing whiff of chordate—our own phylum. The body is laterally compressed and built of some forty segments (a common characteristic of arthropods and several other phyla, including our own). Conway Morris found no hint of the defining arthropod character—jointed appendages. Instead, both the dorsal and ventral (top and bottom) surfaces bear continuous structures that, at least superficially, look like chordate fins supported by fin rays! (With a single specimen, one cannot proceed much beyond the superficial, so this crucial issue remains tantalizingly unresolved.)

Three features of these fins and fin rays deny arthropod affinities and hint chordate: First, a thin and continuous structure, preserved as a dark

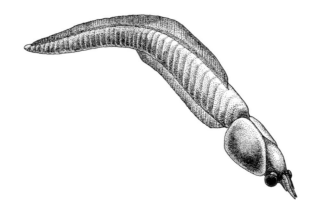

3.28. The enigmatic *Nectocaris,* looking mostly like an arthropod in front and like a chordate with a tail fin behind. Drawn by Marianne Collins.

film on the rock, seems to connect the parallel series of short, stiffening rays into a coherent fin; arthropod limbs, by contrast, are discrete. Second, the fins run along the top and bottom edges of the animal, as in early chordates; arthropod appendages generally attach to the sides of the body. Third, the fins of *Nectocaris* have about three stiffening rays per body division; one pair of appendages per original segment is a defining character of arthropods. (Tagmosis, or coalescence of arthropod segments, is identified by the presence of more than one appendage per division. The segments of *Nectocaris* are too narrow and too numerous for interpretation as amalgamations of several ancestral divisions.)

What can be done with such a chimaera—a creature that looks mostly like an arthropod up front (with possibly unjointed appendages casting some doubt), and mostly like a chordate (or a creature of unknown design) behind? Not much more, when you have but one specimen. So Conway Morris wrote a short, provocative paper and dropped *Nectocaris* into the great holding bin of taxonomy—phylum Uncertain. The title of a taxonomic paper traditionally lists the broad affiliation of the animal being described, but Conway Morris chose a conspicuously noncommittal approach: "*Nectocaris pteryx,* a new organism from the Middle Cambrian Burgess Shale of British Columbia." His final words express no surprise at such a peculiar beast, but hint instead at an emerging generality: "The failure to resolve definitely the affinities of this creature need not be a source of surprise. Current research is showing that a number of species from the Burgess Shale cannot reasonably be accommodated in any extant phylum" (1976a, p. 712).

2. *Odontogriphus.* Conway Morris mounted one rung higher on the ladder of evidence with his second treasure of 1976. He still had only a single specimen, but this time he found both part and counterpart. Walcott had at least set *Nectocaris* aside and supplied a photograph to signal its importance. But *Odontogriphus*—appropriately endowed by Conway Morris with a name meaning "toothed riddle"—was a true discovery, an entirely unnoted specimen, with part and counterpart in separate sections of Walcott's collection. Conway Morris began his paper in the conventional passive voice, but his personal pride and passion come through beneath the stylistic cover-up:

> During a search . . . through the very extensive collection of Burgess Shale fossils . . . a sawn slab bearing the specimen described here was noticed and set aside for further study. Shortly afterwards the counterpart was found elsewhere in the collections. The specimen had evidently never been noted by any other worker. No other specimens have been found (1976b, p. 199).

The fossil of *Odontogriphus* is not well preserved and few structures can be distinguished, but these few are strange indeed. This highly flattened, elongated, oval animal is about two and a half inches long, and marked behind its frontal region with a series of fine, transverse parallel lines, spaced about a millimeter apart. Conway Morris regards these marks as annulations, not separations between true segments. He found no appendages or indications of hardened areas, and assumes that *Odontogriphus* was gelatinous.

The body includes only two resolvable structures, both on the ventral surface at the head end (figure 3.29). A pair of "palps" (probably sensory organs) occupies the corners of the animal's front end. These are shallow rounded depressions formed by up to six platelike layers of tissue parallel to the body surface. The more interesting feature, presumably a mouth surrounded by a feeding apparatus of some kind, lies just forward of the palps, but right in the mid-line. The structure has the form of a shallow, squashed U, opening toward the front. Along the trackway of this U, Conway Morris found some twenty-five "teeth"—tiny pointed, conical structures less than half a millimeter in length. Since these teeth were far too small and fragile to rasp or bite, Conway Morris made the reasonable conjecture that they acted as supports for the bases of tentacles, and that the tentacles, serving as food-gathering devices, surrounded the mouth in a ring.

Such a ring of tentacles would strongly resemble a lophophore—the feeding structure of several modern phyla, notably the bryozoans and brachiopods. Hence, Conway Morris tentatively placed *Odontogriphus* among the so-called lophophorate phyla. But no modern lophophores grow internal teeth to support their tentacles, and nothing else about *Odontogriphus* recalls the form or structure of any other lophophorate animal. "Toothed riddle" remains a fine designation.

Those who follow high-risk strategies must accept the embarrassment of error with the joys of chancy victory. Simon's decision to publish on the rarest and oddest specimens, and to range widely in his interpretations, almost guaranteed some significant mistakes. These come with the territory, and are not badges of dishonor. Simon "made a beauty," as we Yanks used to say, in trying to judge the wider implications of *Odontogriphus*. He couldn't help noticing that its "teeth" bore a vague resemblance to conodonts, then the most enigmatic objects of the fossil record. Conodonts are toothlike structures, often quite complex, that occur abundantly in rocks spanning the great geological range from Cambrian to Triassic (see figure 2.1). They are among the most important of all fossils for geological correlation, but their zoological affinities had long remained mysterious, thus fueling the most famous and long-standing of all paleontological puzzles. Obviously, conodonts are the only hard parts of a soft-bodied animal. But the creature itself had never been found—and what can you tell from some disarticulated teeth?

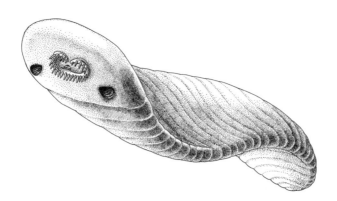

3.29. The flattened swimming animal *Odontogriphus*. The mouth surrounded by tentacles and the pair of palps are shown on the underside of the head. Drawn by Marianne Collins.

Conway Morris thought that the "teeth" of *Odontogriphus* might be conodonts, and that, perhaps, he had discovered the elusive conodont animal. He even took a chance and placed his toothed riddle in the class Conodontophorida. What a potential coup for a beginner—to discover the secret of secrets, and resolve a century of debate! But Simon was wrong. The soft-bodied conodont animal has since been found—with undeniable conodonts lying just in the right place at the forward end of the gut. This creature was also discovered in a museum drawer—in a collection made during the 1920s from a Carboniferous *Lagerstätte* in Scotland known as the Granton Sandstone. The conodont animal, now ranking as one of the few post-Burgess oddballs, looks nothing at all like *Odontogriphus*. Derek Briggs participated in the original description and thinks (though I am not convinced) that the conodont animal may be a chordate, or member of our own phylum (Briggs, Clarkson, and Aldridge, 1983).

3. *Dinomischus.* Simon's third mystery animal carried him another rung up the ladder of evidence. Again, Walcott had set aside and photographed a specimen, but published nothing and left no notes. But this time Conway Morris found himself wallowing in a virtual sea of evidence, for he had three specimens—Walcott's in Washington, another in our collection at Harvard, and a third discovered on Walcott's talus slope by the Royal Ontario Museum in 1975.

All animals discussed so far have been mobile and bilaterally symmetrical. *Dinomischus* represents another major functional design: it is a sessile (fixed and immobile) creature with radial symmetry, suited to receiving food from all directions, like many sponges, corals, and stalked crinoids today. *Dinomischus* looks much like a goblet attached to a long thin stem, with a bulbous holdfast at the bottom to anchor the animal to the substrate (figure 3.30). The entire creature scarcely exceeds an inch in length.

The goblet, called a calyx, bears on its outer rim a series of about twenty elongate, parallel-sided blades, called bracts. The upper surface of the calyx contains both a central and a marginal opening, presumably mouth and anus by analogy with modern creatures of similar habits (figure 3.31). A U-shaped gut, with an expanded stomach at the base, runs between the two openings through the interior of the calyx. Strands radiating from the stomach to the inner surface of the calyx may have been suspensory fibers (for the gut) or muscle bands.

A number of superficial similarities may be noted with bits and pieces of various modern animals, but these are probably broad analogies of similar functional design (like the wings of birds and insects), not detailed homologies of genealogical connection. Conway Morris found closest parallels with a small phylum called the Entoprocta (included with bryozoans in older classifications), but *Dinomischus* is basically a bizarre thing unto itself. Conway Morris showed some hesitation in his original paper (1977a, p. 843), but his latest opinion is unequivocal: *"Dinomischus* has no obvious affinity with other metazoans and presumably belongs to an extinct phylum" (Briggs and Conway Morris, 1986, p. 172).

4. *Amiskwia*. With *Amiskwia*, Simon finally tackled a mainstream Burgess organism, though one of the rarest. Five specimens had been discovered, and Walcott had formally described the genus—as a chaetognath, or arrow worm—in 1911. *Amiskwia* had also been a source of some published debate, though none outside the accepted framework of homes within modern phyla. Two articles in the 1960s had suggested a transfer from the chaetognaths to the nemerteans. These phyla are not household names, but both are staples of modern taxonomy.

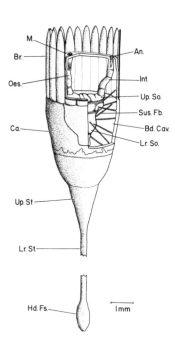

3.30.   Original reconstruction of *Dinomischus* by Conway Morris (1977a). Part of the calyx is broken away to show the interior anatomy of the organism. Note the U-shaped gut going from the mouth (labeled *M.*) to the anus *(An.),* and the muscle bands *(Sus. Fb.,* for "suspensory fibers") anchoring the gut to the wall of the calyx.

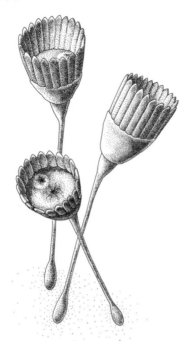

3.31. Three specimens of the stalked animal *Dinomischus.* One bends toward us, showing the openings of the mouth and anus on the top of the calyx. Drawn by Marianne Collins.

Amiskwia, as a compressed, probably gelatinous animal with no outer carapace, did squash flat on the Burgess rock surfaces. Hence, these fossils are truly preserved in the mode that Walcott incorrectly viewed as normal for all Burgess organisms—as a flat sheet. Without the three-dimensional structure that Whittington found for arthropods, and that Simon confirmed for several other oddballs, little of *Amiskwia's* anatomy can be well resolved—though enough has been preserved to preclude a place in any modern phylum.

The head region bears a pair of tentacles, inserted on the front ventral surface (figure 3.32). The trunk sports two fins, unsupported by rays or any other stiffening device, in the plane of body flattening—lateral (at the sides) and caudal (forming a tail). (The chaetognaths often have fins in roughly similar positions, hence Walcott's designation. But a true chaetognath also has a head with teeth, hooks, and a prominent hood—and no tentacles. Nothing else about *Amiskwia* even vaguely suggests chaetognath affinities, and the rough similarity of fins represents separate evolution for similar function in swimming.) *Amiskwia* is probably one of the few

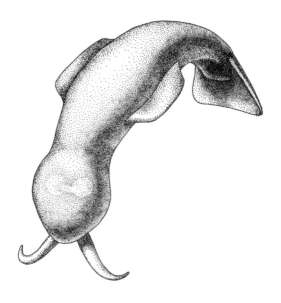

3.32. The flattened swimming animal *Amiskwia*, with a pair of tentacles on the head, and side and tail fins behind. Drawn by Marianne Collins.

Burgess animals that did not live in the bottom community engulfed by the mudslide. It was presumably a pelagic (or swimming) organism, living in open waters above the stagnant basin that received the Burgess mudslide. This different mode of life would explain the great rarity of *Amiskwia, Odontogriphus,* and a few other creatures that may have lived in open waters above the grave, but away from the original home, of the main Burgess community. Only a few animals of the water column above would have died and settled into the sediments below during the short time when the mudslide was coalescing into a layer of sediment in the stagnant basin.

Within the head, a bilobed organ may represent cerebral ganglia, while the gut can be traced as a straight tube from an enlarged region at the head to an anus at the other end of the body, just in front of the caudal fin (figure 3.33). The head, lacking the characteristic proboscis with a prominent fluid-filled cavity and muscular walls, looks nothing like that of a nemertean—the other candidate for a conventional taxonomic home; while the caudal fin exhibits only superficial similarity (in nemerteans, the fin is bilobed, and the anus opens at the very tip of the body). Conway Morris, now becoming quite comfortable with the idea of taxonomic uniqueness at high anatomical levels, concluded:

While *Amiskwia sagittiformis* is certainly not a chaetognath, ... the worm cannot be placed within the nemerteans either. The relative similarity ... [to nemerteans] is regarded as superficial and merely a product of parallel evolution. *Amiskwia sagittiformis* does not appear to be more closely related to any other known phylum (1977b, p. 281).

5. *Hallucigenia.* We need symbols to represent a diversity that we cannot fully carry in our heads. If one creature must be selected to bear the message of the Burgess Shale—the stunning disparity and uniqueness of anatomy generated so early and so quickly in the history of modern multicellular life—the overwhelming choice among aficionados would surely be *Hallucigenia* (though I might hold out for *Opabinia* or *Anomalocaris*). This genus would win the vote for two reasons. First, to borrow today's vernacular, it is really weird. Second, since names matter so much when we

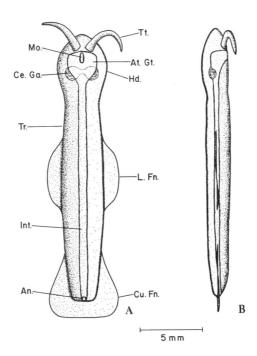

3.33. Reconstruction of *Amiskwia* by Conway Morris (1977b). (A) Bottom view: note the insertion of the tentacles (labeled *Tt.*), the position of the mouth *(Mo.)*, the path of the gut *(Int.)* to the anus *(An.)*, and the structure interpreted as possible cerebral ganglia *(Ce. Ga.)*. (B) Side view.

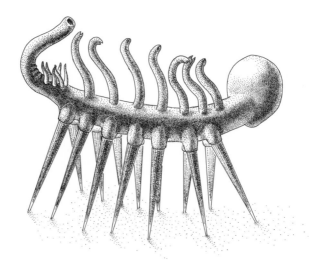

3.34. *Hallucigenia,* supported by its seven pairs of struts, stands on the sea floor. Drawn by Marianne Collins.

are talking about symbols, Simon chose a most unusual and truly lovely designation for his strangest discovery. He called this creature *Hallucigenia* to honor "the bizarre and dream-like appearance of the animal" (1977c, p. 624), and also, perhaps, as a memorial to an unlamented age of social experiment.

Walcott had assigned seven Burgess species to *Canadia,* his principal genus of polychaetes. (Polychaetes, members of the phylum Annelida, the segmented worms, are the marine equivalent of terrestrial earthworms, and are among the most varied and successful of all animal groups.) Conway Morris later showed (1979) that Walcott's single genus was hiding remarkable disparity under one vastly overextended umbrella—for he eventually recognized, among Walcott's seven "species," three separate genera of true polychaetes, a worm of an entirely different phylum (a priapulid that he renamed *Lecythioscopa*), and *Hallucigenia.* Walcott, mistaking the strangest of all Burgess creatures for an ordinary worm, referred to this oddball as *Canadia sparsa.*

How can you describe an animal when you don't even know which side is up, which end front and which back? *Hallucigenia* is bilaterally symmetrical, like most mobile animals, and carries sets of repeated structures in common with the standard design of many phyla. The largest specimens are about an inch long. Beyond these vaguest of familiar signposts, we are forced to enter a truly lost world (figure 3.34). In broad outline, *Hallucigenia* has a bulbous "head" on one end, poorly preserved in all available specimens (about thirty), and therefore not well resolved. We cannot even be certain that this structure represents the front of the animal; it is a "head" by convention only. This "head" (figure 3.35) attaches to a long, narrow, basically cylindrical trunk.

Seven pairs of sharply pointed spines—not jointed, arthropod-like appendages, but single discrete structures—connect to the sides of the trunk, near the bottom surface, and extend downward to form a series of struts. These spines do not articulate to the body, but seem to be embedded within the body wall, which extends as a sheath for a short distance along the top of each spine. Along the dorsal mid-line of the body, directly opposite the spines, seven tentacles with two-pronged tips extend upward. The seven tentacles seem to be coordinated with the seven pairs of spines in an oddly displaced but consistent way: the first tentacle (nearest the "head") corresponds to no spine below. Each of the next six tentacles lies directly above a pair of spines. The last pair of spines has no corresponding tentacle above. A cluster of six much shorter dorsal tentacles (perhaps arranged as three pairs) lies just behind the main row of seven. The posterior end of the trunk then narrows into a tube and bends upward and forward.

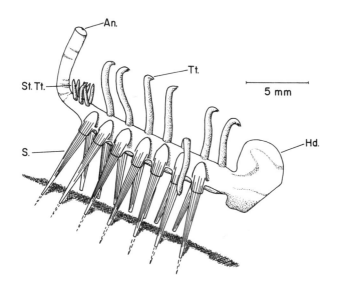

3.35. Original reconstruction of *Hallucigenia* by Conway Morris (1977c).

How can a taxonomist proceed in interpreting such a design? Simon decided that he must first try to figure out how such an animal could operate; then he might gain some further clues to its anatomy. Searching for analogies, Simon noted that some modern animals rest upon, and even move with, spines attached to their bottom sides. "Tripod" fish support themselves upon two long pectoral spines and one tail spine. The elasipods, a curious group of deep-sea holothurians (sea cucumbers of the echinoderm phylum), move in groups along the bottom, supported by elongate, spiny tube feet (Briggs and Conway Morris, 1986, p. 173). In *Hallucigenia*, the two spines of each pair meet at an angle of some seventy degrees, an excellent arrangement for a series of struts supporting the body in fair stability. Conway Morris therefore began by supposing that the seven pairs of spines permitted *Hallucigenia* to rest on a muddy substrate. This assumption defines both a mode of life and an orientation: "Dorsal and ventral surfaces are identified on the assumption that the spines were embedded in the bottom sediments" (Conway Morris, 1977c, p. 625).

So far, so good; *Hallucigenia* could rest on the bottom in fair stability. But the animal couldn't stand there in perpetuity like a statue; bilaterally symmetrical creatures with heads and tails are almost always mobile. They concentrate sensory organs up front, and put their anuses behind, because they need to know where they are going and to move away from what they leave behind. How in heaven's name could *Hallucigenia* move on a set of spikes fixed firmly into the body wall? Conway Morris did manage to suggest a plausible model, in which strips and bands of muscle anchor the proximal end of the spine to the inner surface of the body wall. Differential expansion and contraction of these bands could move the spines forward and back. A coordinated wave of such motion along the seven pairs might propel the animal, if a bit clumsily. He was not thrilled with the prospects for such a mode of locomotion, and suggested that *"Hallucigenia sparsa* probably did not progress rapidly over rocks or mud, and much of its time may have been spent stationary" (1977c, p. 634).

If the spines are hard to interpret, what about the tentacles above—where prospects for modern analogues are dimmer. The pincers at their tips could have captured food, but the tentacles don't reach the head region, and passage of food from one tentacle to another toward a frontal mouth offers little promise of efficient eating. Noting a possible connection between a hollow tube within each tentacle and a gut within the trunk (neither well enough preserved to inspire confidence), Conway Morris offered a fascinating alternative. Perhaps *Hallucigenia* had no frontal mouth at all. Perhaps each tentacle gathered food independently, passing the collected particles down its own personal gullet into the communal gut.

You have to consider bizarre solutions when you work with such a strange animal.

Yet *Hallucigenia* is so peculiar, so hard to imagine as an efficiently working beast, that we must entertain the possibility of a very different solution. Perhaps *Hallucigenia* is not a complete animal, but a complex appendage of a larger creature, still undiscovered. The "head" end of *Hallucigenia* is no more than an incoherent blob in all known fossils. Perhaps it is no head at all, but a point of fracture, where an appendage (called *Hallucigenia*) broke off from a larger main body (yet undiscovered). This prospect may seem disappointing, since *Hallucigenia* by itself forms such a wondrous beast. Hence, I am rooting for Conway Morris's interpretation (but if forced to bet, I would have to place my money on the appendage theory). But then, the prospect of *Hallucigenia* as only an appendage may be even more exciting—for the whole animal, if ever discovered and reconstructed, might be even more peculiar than *Hallucigenia* as now interpreted. It has happened before in the Burgess. *Anomalocaris* (see Act 5) was once viewed as an entire arthropod, and a fairly dull crustacean at that. Then Whittington and Briggs (1985) resolved it as a feeding appendage of an animal ranking just behind *Hallucigenia* in Burgess oddity. We have surely not seen the last, and perhaps not the greatest, of Burgess surprises.

DEREK BRIGGS AND BIVALVED ARTHROPODS: THE NOT-SO-FLASHY BUT JUST-AS-NECESSARY FINAL PIECE

I must begin with an apology to Derek Briggs for an invisible slight arising from both ignorance and thoughtlessness. I made a bad mistake when I first laid out this chronological centerpiece of the book—that is, before I read the monographs in detail. I saw the Burgess transformation as a dramatic interplay between Harry Whittington, the conservative systematist who started it all, and Simon Conway Morris, the young and radical man of ideas who developed a revolutionary interpretation and dragged everyone else along. I have already indicated my error in reading this interaction according to the conventional script.

Let me now confess another mistake, one that I should not have made. This is the classic error of those who write about science without an intuitive feel for its daily procedures; those who do the work should know better. The journalistic tradition so exalts novelty and flashy discovery, as reportable and newsworthy, that standard accounts for the public not only miss the usual activity of science but also, and more unfortunately, convey a false impression about what drives research.*

A project like the Burgess revision has potentially flashy and predictably less noticeable aspects. Both are necessary. A conventional reporter will convey only the hot ideas and the startling facts—*Hallucigenia* gets ink; the Burgess trilobites get ignored. But the Burgess oddballs mean little in isolation. When placed in an entire fauna, filled with conventional elements as well, they suggest a new view of life. The conventional creatures must be documented with just as much love, and just as assiduously—for they are every bit as important to the total picture.

Derek Briggs drew the bivalved arthropods as his subject—the apparently most conventional group in the Burgess fauna. He produced an elegant series of monographs on these animals, finding some surprises, but also confirming some expectations. I had not appreciated the central role that Briggs's work on bivalved arthropods played in the Burgess transformation. As I read Derek's monographs, I recognized my error with some shame, and grew to understand Harry, Derek, and Simon as a trio of equals, each with a distinct and necessary role in the total drama.

Walcott and others had described about a dozen genera of arthropods with a bivalved carapace (usually enclosing the entire head and front part of the body). Several of these genera cannot be classified with certainty, for only the carapaces have been found, not the soft parts. The other genera have always, and without any doubt or hesitation, been identified as crustaceans—as are all modern arthropods with a bivalved carapace. Derek Briggs began his project without any conscious doubts: "There were some redescriptions to be done. I assumed I would be dealing with a bunch of crustaceans."

Briggs described two outstanding discoveries in his first monographs on the bivalved arthropods of the Burgess Shale. Put these together with Simon's oddballs and Harry's orphaned arthropods, and you have, by 1978, both a fully articulated and completely new account of how multicellular animal life evolved.

1. *Branchiocaris,* the first discovery. The Crustacea are an enormous and diverse group—from the nearly microscopic ostracodes with bivalved carapaces covering the entire body like a clamshell, to giant crabs with leg spreads of several feet. Yet all are built upon a stereotyped ground plan, with a definite signature in the structure of the head. The crustacean head

---

*I don't say this in a critical, revelatory, or muckraking mood. Journalistic traditions properly match their assigned roles. I only point out that different approaches see only restricted parts of a totality—as in the overworked simile of the blind men and the elephant—and that one can get something gloriously wrong by mistaking a small and biased segment for an entity.

is an amalgam of five original segments plus eyes. Five pairs of appendages are therefore present—and in a definite arrangement: two pre-oral (usually antennae) and three post-oral (usually mouth parts).* Since all modern bivalved arthropods are crustaceans, Briggs assumed that he would find this frontal signature in his Burgess subjects. But the Burgess soon provided yet another surprise.

Back in 1929, Charles E. Resser, Walcott's right-hand man at the Smithsonian, had described a single Burgess specimen as the crustacean *Protocaris pretiosa.* The genus *Protocaris* had been established in 1884, by none other than Charles Doolittle Walcott in his pre-Burgess days, for a Cambrian arthropod from the Parker Slate of Vermont. Resser considered the Burgess animal as sufficiently close for inclusion in the same genus. Briggs disagreed and established the new genus *Branchiocaris.*

Briggs managed to amass a total of five specimens—Resser's original, three more from the Walcott collection, and a fifth whose part was found by Raymond in 1930, but whose counterpart remained on the Burgess talus until collected by the Royal Ontario Museum expedition in 1975, as recounted in the heart-warming tale earlier in this chapter. The bivalved carapace of *Branchiocaris* covers the head and anterior two-thirds of the body (figure 3.36). The body itself contains some forty-six short segments, with a two-pronged telson behind. The appendages are not clearly distinguishable in the limited number of available fossils, but may have been biramous, with a short segmented branch (presumably homologous to the walking leg of most biramous arthropods, but too reduced for such a function in *Branchiocaris*), and a larger bladelike process, probably used for swimming near the sea floor.

But the head of *Branchiocaris* provided the big surprise. Two pairs of short antenna-like appendages, pointing forward, could clearly be seen— the first more conventional in form, uniramous with many segments; the second more peculiar, stout and composed of few segments, perhaps with a claw or pincer at the end. Briggs called this second pair the "principal appendage"— just as Whittington, stumped by an analogous structure in *Yohoia,* had spoken of a "great appendage."

These appendages attached to the upper and lateral surfaces of the head. On the ventral side, three pairs of additional appendages should have

---

*The mouth parts of arthropods have been given the same names as functionally comparable structures in vertebrates—maxilla, mandible, and so forth. Similarly, the parts of insect legs bear the same names—trochanter, tibia—as their vertebrate counterparts. This is an unfortunately confusing nomenclature, for whatever the functional similarities, the structures have no evolutionary connection: insect mouth parts evolved from legs; vertebrate jaws from gill arches.

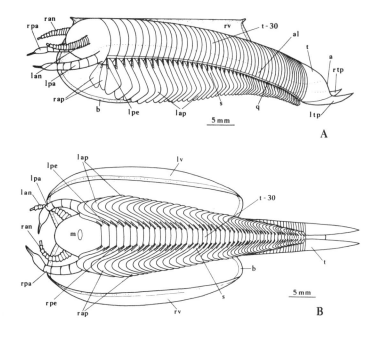

3.36.   Reconstruction of *Branchiocaris* by Briggs (1976). (A) Side view.
(B) Bottom view, showing the ventral surface of the animal surrounded by
the two valves of its carapace. Note in particular the pairs of uniramous
appendages, especially the unique principal appendage (labeled *lpa* and *rpa*).
And note also the absence of any appendages on the head behind the mouth;
this arrangement is unknown in any modern arthropod group.

followed the mouth. Briggs found nothing. The mouth stood all alone on
an unadorned ventral surface. *Branchiocaris*, with two and only two pairs
of appendages on the head, was not a crustacean. "It apparently defies
classification within any group of Recent arthropods," Briggs concluded
(1976, p. 13).

Thus, the bivalved arthropods—the group that seemed most promising
as a coherent set of evolutionary cousins—also formed an artificial category
hiding an unanticipated anatomical disparity. What order could possibly
be found among the Burgess arthropods? Each one seemed to be built
from a grabbag of characters—as though the Burgess architect owned a
sack of all possible arthropod structures, and reached in at random to pick
one variation upon each necessary part whenever he wanted to build a new
creature. Could a biramous limb of trilobite type adorn any kind of arthro-

pod body? Could a bivalved carapace cover any anatomy? Where was order, where decorum?

2. *Canadaspis*, the second discovery. Consider the story of Burgess arthropods as published by the end of 1976. *Marrella*, a supposed relative of trilobites, was an orphan. *Yohoia*, with its great appendage, was uniquely specialized and unaffiliated, not a precursor of anything. *Burgessia*, namesake of the fauna, was another orphan. Even *Branchiocaris*, firm candidate for a crustacean, sported a unique anatomy under its bivalved carapace. Moreover, these four orphans showed no propensity for coherence among themselves; each reveled in its own peculiarities. Would any Burgess arthropod ever accept the allegiance to a modern group that Walcott, wielding his shoehorn, had once forced upon all?

Canadaspis is the second most common animal in the Burgess Shale. It is large by Burgess standards (up to three inches in length) and tends to be preserved with a conspicuous reddish color. It has a bivalved carapace, but as Briggs soon discovered, an underlying anatomy very different from *Branchiocaris*.

In a short paper of 1977, Briggs placed two bivalved species in the new genus, *Perspicaris*. His reconstructions suggested something exciting, but the rarity of specimens and their poor preservation precluded any firm conclusion. He couldn't prove the affiliation, but nothing about these two species precluded membership in the Crustacea. Had a representative of a modern group finally been found?

In 1978, Briggs resolved this issue with elegance and finality. His long

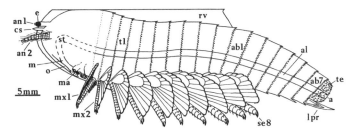

3.37. Reconstruction of *Canadaspis* by Briggs (1978). This animal has the typical structure of a true crustacean of the malacostracan line: two pairs of appendages in front of the mouth (labeled *an1* and *an2*), three pairs of appendages behind the mouth *(ma, mx1,* and *mx2)*, a thorax of eight segments (beginning with the segment labeled *t1*), and an abdomen of seven segments *(ab1–ab7).* Each thoracic segment bears a pair of biramous appendages.

monograph on the well-preserved, superabundant *Canadaspis perfecta* finally placed a Burgess creature in a successful modern group. *Canadaspis* was not only a crustacean, but its home within the Crustacea could be established. *Canadaspis* is an early malacostracan—a representative of the great group of crabs, shrimp, and lobsters. Briggs found all elements of the intricate malacostracan stereotype in the anatomy of *Canadaspis:* a head bearing five pairs of appendages, and built of five segments plus eyes; a thorax (middle section) of eight segments, and an abdomen (back section) of seven segments plus a telson. Further, the head appendages are arranged just right, with two pairs of short, uniramous antennae in front of the mouth, and three pairs of ventral appendages behind the mouth.* The abdominal segments bear no appendages, but each thoracic segment carries a pair of standard biramous appendages, with an inner leg branch and a broad outer gill branch (figures 3.37 and 3.38).

The brevity of this description is no denigration of the importance of *Canadaspis* in the Burgess reformulation. A weird animal needs a longer write-up to explain its uniqueness; a familiar creature can simply be characterized as "like Joe whom everyone knows." But *Canadaspis* is both a key and an anchor to the Burgess story, a creature every bit as important as any of Simon's weird wonders. Suppose that every Burgess animal were a bizarre denizen of a lost world. What then would we make of the assemblage? A failed experiment, a washout, a first attempt totally bypassed by a reconstituted modern fauna, and therefore offering no clues and no connection to the origin of later life. But the presence of *Canadaspis,* and

---

*As an indication of how much struggle and effort can underlie the conclusions stated so briefly in my text, consider this interesting note that Derek Briggs wrote to me as a reaction to this passage when I sent the manuscript of this book to him: "The work on *Canadaspis* became a hunt for the first crustacean. . . . By then the expectation was that the odds on any of the arthropods falling into living groups were very low. The problem with *Canadaspis* was finding the critical evidence of the posterior cephalic [head] appendages. USNM 189017 [catalog number of a key specimen in the United States National Museum] is the best of only about 3 (out of thousands) specimens which show these limbs in lateral view (they are almost without exception totally obscured by the carapace, compaction etc.), and as you can see on Plate 5 (Briggs 1978) it was a huge job preparing the specimen to show them. In my view figs. 66–69 on that plate represent the peak of what can be achieved by preparing part and counterpart in tandem. I then had a major job convincing Sidnie Manton (Harry's arthropod guru) that I did indeed have the critical evidence—at the time I considered that an enormous achievement! [Manton was the world's greatest expert on the higher-level classification of arthropods—and one tough lady.] It was not just a case of the evidence of the specimens; it was necessary to argue that the first two pairs of a series of 10 pairs of similar biramous appendages belonged to the head—although they remain primitive in not being significantly differentiated from those which follow."

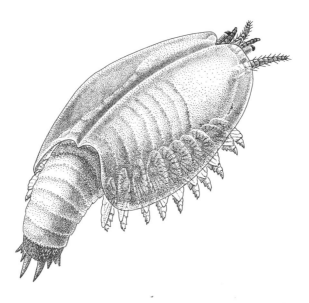

3.38. The true crustacean *Canadaspis*. The five head segments bear two pairs of antennae and three pairs of appendages behind the mouth, the last two of which are continuous with, and similar in form to, the biramous appendages of the body. Drawn by Marianne Collins.

other creatures of modern design, suggests a different and more enlightening view. The Burgess fauna does include modern prototypes, and, in this key respect is an ordinary Cambrian fauna; but the vastly broader range of designs that disappeared may reveal the most important of all patterns in life's early history.

As Derek resolved *Canadaspis*, Simon had left behind his whirlwind of wonders to work on the main subjects of his project, the true Burgess worms. His results, published in two monographs (1977 and 1979), beautifully affirmed the lesson of *Canadaspis*. Some Burgess organisms, even among soft-bodied members of the fauna, fit comfortably into modern groups—thus accentuating and highlighting the importance of the oddballs as additions to normality. In 1977, Conway Morris recognized among forms that Walcott had scattered across three phyla (as polychaetes, crustaceans, and echinoderms) six or seven genera of priapulid worms. The Priapulida form a small phylum of ten genera or so in today's oceans, but they dominated the worm fauna of the Burgess Shale. (The Burgess priapulids form a major part of my story in chapter V.)

In 1979, Conway Morris sorted out one of Walcott's greatest confusions—the Burgess polychaetes. Walcott had used the Polychaeta (marine representatives of the phylum Annelida, or segmented worms) as a dumping ground for many Burgess oddities. Within Walcott's polychaetes, Conway Morris found two genera of priapulids and four genera of weird wonders. But Walcott had also identified some true polychaetes. From this mixture, Conway Morris identified and established six genera of Burgess polychaetes. This group, so dominant in today's seas, was overshadowed by priapulids (with the same number of genera, but many more specimens) in Burgess times. But both groups proclaimed the same general message. The Burgess fauna contained both ordinary and unique anatomies in abundance.

ACT 4. Completion and Codification of an Argument: *Naraoia* and *Aysheaia*, 1977–1978

After such an extended third act, we need a sparer fourth to make a largely symbolic point amidst the resolution of two important Burgess genera distinguished by more than their maximally unpronounceable, vowel-laden names.

Harry Whittington had started this drama by orphaning some arthropods that everyone had previously placed in established groups (Act 1). He had upped the ante by showing that *Opabinia* was not an arthropod at all, but a creature of strange and unique anatomy (Act 2). His students and associates then converted these anomalies into a generality about the Burgess and its time by documenting the same pattern throughout the fauna (Act 3). When Harry Whittington finally accepted the new interpretation, and began to view anatomical oddity as a preferred hypothesis *a priori*, rather than a last resort, the story had reached its logical end; the Burgess transformation had been completed (Act 4). In conceptual terms, the rest would be mopping up, but with the best of all particular stories still to be told (Act 5).

Naraoia added the last substantial piece to the logical structure of the new view. This old Burgess standby, described by Walcott as a branchiopod crustacean, has a carapace composed of two flat, smooth, oval valves, meeting at straightened borders one behind the other. These valves, discrete and shiny on most fossils, make *Naraoia* one of the most striking and attractive of Burgess organisms, but they also impose a severe problem in interpretation. They cover almost all the soft anatomy; most specimens show only the distal tips of the appendages, protruding out beyond the edge of the carapace (figure 3.39). Since the proximal (and invisible) ends of the appendages provide the primary taxonomic basis for identifying arthropod groups—both by their form and by their pattern of insertion into the body—*Naraoia* could never be properly interpreted.

Whittington resolved this dilemma with his discovery of three-dimensional structure in the Burgess fossils. He realized that he could dissect through the firm carapace to reveal the proximal ends of the appendages, and their points of insertion. When he cut through the carapace of *Naraoia* (figure 3.40), he uncovered enough of the appendages to count their segments and resolve their proximal ends, including gnathobases and food grooves. Whittington also received one of the great surprises of his professional life. He was looking at a leg branch of the animal he knew best—a trilobite. But beyond a vague similarity in general outline, the carapace, with its two valves, hardly resembles the exoskeleton of a trilobite. Most trilobites have a threefold division, into head, thorax, and pygidium. (Contrary to popular belief, this division, stem to stern, is not the source of the name "trilobite," or "three-lobed." Trilobation refers to

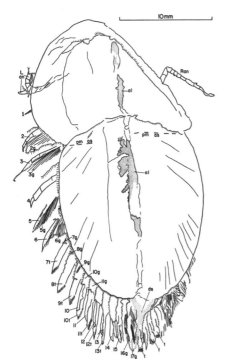

3.39.   Camera lucida drawing of an excellent specimen of *Naraoia* (Whittington, 1977). The two valves of the carapace cover almost all the soft anatomy, and only the ends of the appendages protrude beyond them.

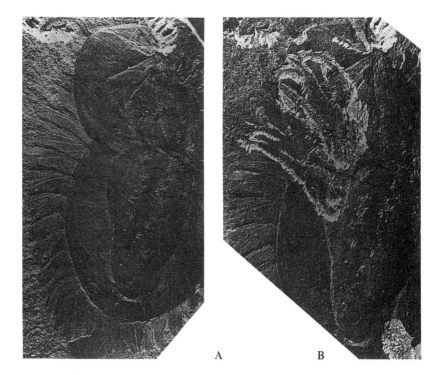

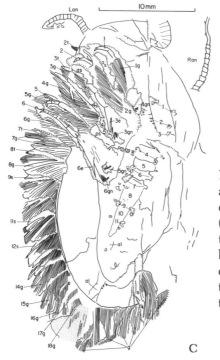

A   B

C

Lon   10mm
Ran

3.40. Determination of the taxonomic affinity of *Naraoia* by dissection. (A) A complete specimen before dissection. (B) The same specimen, dissected to reveal the legs at their point of attachment to the body. (C) Camera lucida drawing of the dissected specimen. Since the legs are of typical trilobite form, *Naraoia* is identified as the first known bivalved trilobite.

the threefold side-to-side division into a central axis and two side regions, called pleurae.)

Whittington also found other key trilobite characters in *Naraoia,* notably the defining segmentation of the head, with one pair of uniramous pre-oral antennae and three pairs of ventral post-oral appendages. *Naraoia,* despite its curious outer covering, was surely a trilobite. Whittington therefore described this genus as a new and separate class within the Trilobita. He wrote with barely disguised joy and an uncharacteristic personal touch—and why not, for Harry is the world's expert on trilobites. These are his babies, and he had just given birth to a stunning and different child:

> It was both surprising and exciting to excavate for the first time. . . . The new reconstruction shows a very different animal from Walcott's and other restorations, . . . far more trilobite like than had been thought. Indeed, I conclude *Naraoia* was a trilobite that lacked a thorax, and place it in a separate order of that class (1977, p. 411).

This change may seem small, a shift from one well-known group to another, and therefore an event of little conceptual interest in the midst of so much Burgess turmoil and discovery. Not so. The classification of *Naraoia* is a satisfying final piece of a puzzle, proving that the basic Burgess pattern—anatomical disparity beyond the range of later times—applies at all levels. Simon's weird wonders had established the pattern at the highest level of phyla, the basic ground plans of animal life. Whittington's monographs had told the same story at the next lower level of disparity within phyla—group after group of orphaned arthropods spoke of Burgess anatomy far beyond the range of any later time, despite a vast increase in arthropod species, including a modern insect fauna of nearly a million described species. Now Harry had demonstrated the same pattern again at the lowest level of disparity within major groups of a phylum. He had discovered an apparent contradiction in terms—a soft-bodied trilobite with a carapace of two valves. (In 1985 he would describe a second soft-bodied trilobite, *Tegopelte gigas,* one of the largest Burgess animals at nearly a foot in length, so *Naraoia* is no lone oddity among trilobites.) The Burgess pattern seems to display a "fractal" character of invariance over taxonomic scales: crank up the telescope, or peer down the microscope, and you see the same picture—more Burgess disparity, followed by decimation and diversification within fewer surviving groups.

The monograph on *Naraoia* marked a conceptual watershed for Whittington. He finally sank the class Trilobitoidea officially, as an artificial wastebasket with no evolutionary validity. He had finally freed himself to view the Burgess arthropods as a series of unique designs, beyond the range of later groups:

> The Class Trilobitoidea Størmer, 1959 was proposed as a convenient category in which to place various supposedly trilobite-like arthropods, mainly from the Burgess Shale, and regarded as of equal rank to the Class Trilobita. Studies recently published and in progress are providing abundant new information, particularly on appendages. . . . The Class Trilobitoidea can no longer be regarded as a useful concept, and a new basis for assessment of relationships is emerging (1977, p. 440).

Harry's next monograph, on *Aysheaia*, begins with his most explicit recognition of the new view: "The animals in this community include an astonishing variety of arthropods as well as bizarre forms, such as those described by Whittington and Conway Morris which, like *Aysheaia*, are not readily placed in Recent higher taxa" (1978, pp. 166–67). *Aysheaia* was perhaps the most famous and most widely discussed of Burgess organisms—for an interesting reason rooted in the two *p*'s, "primitive" and "precursor." Walcott (1911c) had described *Aysheaia* as an annelid worm, but colleagues soon pointed out with excitement that the creature could hardly be distinguished, at least superficially, from a small group of modern invertebrates called the Onychophora and represented primarily by a genus with the lovely name *Peripatus*. The Onychophora possess a mixture of characters recalling both annelids and arthropods; many biologists therefore regard this group as one of the rare connecting forms ("nonmissing links," if you will) between two phyla. But modern Onychophora are terrestrial, while the actual transition from annelid to arthropod, or the derivation of both from a common ancestor must have occurred in the sea. In addition, modern Onychophora have undergone more than 550 million years of evolution since the supposed linkage of annelid and arthropod, and could not be viewed as direct models of the transition. A marine onychophoran from the Cambrian would be a creature of supreme evolutionary importance—and *Aysheaia*, generally so interpreted (Hutchinson, 1931), became a hero of the Burgess. The great ecologist G. Evelyn Hutchinson, who had done important work on the taxonomy of *Peripatus* in South Africa, and who, looking back on a rich career from his ninth decade, still places his study of *Aysheaia* among his most significant (interview of April 1988), wrote:

> In *Aysheaia* we have a form living under entirely different ecological conditions from those of the modern species, and at a very remote time, yet having

an external appearance, which in life must have been extraordinarily similar to that of the living representatives of the group (1931, p. 18).

Aysheaia has an annulated, cylindrical trunk, with ten pairs of annulated limbs attached at the sides near the lower surface, and pointing down, presumably for use in locomotion (figures 3.41 and 3.42). The anterior end is not separated as a distinct head. It bears a single pair of appendages, much like the others in form and annulation but attached higher on the sides and pointing laterally. The terminal mouth (smack in the middle of the front surface) is surrounded by six or seven papillae. The head appendages bear three spinelike branches at their tip, and three additional spines along the anterior margin. The body limbs end in a blunt tip carrying a group of up to seven tiny, curved claws. Larger spines emerge from the limbs themselves. These spines are absent on the first pair, point forward on pairs 2–8, and backward on 9–10.

Whittington combined this anatomical information with other data to reconstruct an interesting and unusual life style for *Aysheaia*. On or near six of the nineteen *Aysheaia* specimens he found remains of sponges—an association hardly ever encountered with other Burgess animals. Whittington conjectured that *Aysheaia* might have fed on sponges and lived among them for protection as well (figure 3.43). The tiny terminal claws of the limbs would not have worked on mud, but might have helped in climbing upon sponges and holding on. The anterior appendages could not have

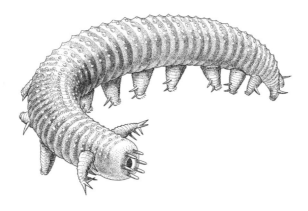

3.41. *Aysheaia,* probably an onychophoran. Drawn by Marianne Collins.

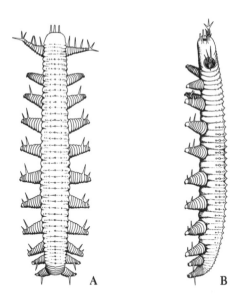

3.42. Reconstruction of *Aysheaia* by Whittington (1978). (A) Top view.
(B) Side view: the ring of tentacles surrounding the terminal mouth is visible
at the top; the dorsal surface is at the right.

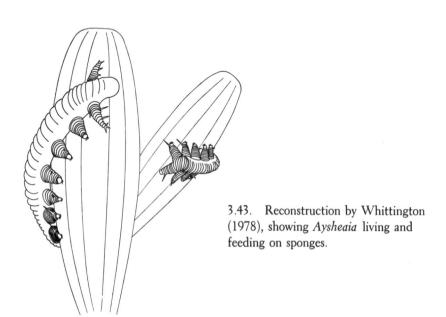

3.43. Reconstruction by Whittington
(1978), showing *Aysheaia* living and
feeding on sponges.

swept food directly into the mouth, but they might have lacerated sponges with their spines, permitting the animal to lap up nutritious juices and soft tissues. The backward-facing claws and spines of the posterior body limbs might have functioned as anchors to keep the animal in place at odd angles.

But was *Aysheaia* an onychophoran? Whittington admitted some impressive similarities in the anterior appendages, the short, uniramous body limbs with terminal claws, and the annulations on body and limbs. But he also cited some differences, including lack of jaws (possessed by modern onychophorans) and the termination of the body at the last pair of limbs (the body extends farther back in modern onychophorans).

In Whittington's judgment, these differences raised sufficient doubts to debar *Aysheaia* from the Onychophora and to recognize this genus, albeit tentatively, as a unique and independent group. Citing the lessons of other genera, he wrote: "Thus *Aysheaia*, like other Burgess Shale animals as *Opabinia*, *Hallucigenia*, and *Dinomischus*, does not fit readily into any extant higher taxon" (1978, p. 195).

I regard these words as momentous, and (symbolically, at least) as the completion of the Burgess transformation. I say this, ironically, because I think that for once, Harry was probably wrong about *Aysheaia*. I believe that, on the balance of evidence, *Aysheaia* should be retained among the Onychophora. The similarities are impressive and anatomically deep, the differences superficial and not of great evolutionary import. Of Harry's two major differences, jaws may simply have evolved later. Structures can be added in evolution provided that ancestral anatomies do not preclude their development. Just such an event occurred in at least one prominent Burgess group. Burgess polychaetes have no jaws, but jaws evolved by Ordovician times and have persisted ever since. As for the extension of the body beyond the last pair of limbs, this strikes me as an easy evolutionary change, well within the capacity of a broad group like the Onychophora. The American paleontologist Richard Robison, who developed a much longer list of distinctions between *Aysheaia* and modern onychophorans, agrees nonetheless that *Aysheaia* belongs in the group, and writes of Whittington's second major difference:

> In terrestrial onychophorans, projection of the body behind the posterior pair of lobopods [limbs] seems to represent nothing more than minor modification to improve sanitation by slight displacement of the anus. Such body design is less important to animals living in water, where currents aid separation of toxic waste from the body. Thus, posterior shape of the body may be more indicative of habitat than phylogenetic affinity (1985, p. 227).

Why then did Whittington separate *Aysheaia* from the Onychophora and assert its taxonomic uniqueness? Since this conclusion came from a man who, for years, had been resisting the temptation to separate Burgess organisms from well-known groups, and who had made such divisions only when forced by weight of evidence, we would naturally assume that he had been compelled to this uncomfortable conclusion by new data direct from *Aysheaia*. But read the 1978 monograph carefully. Whittington did not upset any of Hutchinson's basic statements about *Aysheaia*. Harry had listed and discussed the same differences; he had essentially affirmed, in much greater and more elegant detail to be sure, Hutchinson's excellent work. But Hutchinson had classified *Aysheaia* as an onychophoran—on the very same data that Whittington later used to reach the opposite conclusion.

What then had prompted Whittington's reversal, if not the anatomy of *Aysheaia?* We have a reasonably well-controlled psychological experiment here. The data had not changed, so the reversal of opinion can only record a revised presupposition about the most likely status of Burgess organisms. Obviously, Whittington had come to accept, and even to prefer, the idea of taxonomic uniqueness for animals of the Burgess Shale. His conversion was complete.

Many fascinating genera still awaited description; the halfway point had not even been reached. But Whittington's 1978 monograph on *Aysheaia* marks the codification of a new view of life. What a dizzying few years between 1975 and 1978—from the disturbing discovery that *Opabinia* is neither an arthropod, nor anything else ever known before, through the cascade of Simon's weird wonders, to the full acceptance of taxonomic uniqueness as a preferred hypothesis. Three short years and a new world!

ACT 5. The Maturation of a Research Program:
Life after *Aysheaia*, 1979–Doomsday
(There Are No Final Answers)

The seven short years from *Marrella* (1971) to *Aysheaia* (1978) had produced an extraordinary shift of perspective—from a project designed to redescribe some arthropods classified in familiar groups, to a new conception of the Burgess Shale and the history of life.

The pathway had not been smooth and direct, clearly marked by the weight of evidence and logic of argument. Intellectual transformations never proceed so simply. The flow of interpretation had meandered and backtracked, mired itself for a time in a variety of abandoned hypotheses (on the primitive status of Burgess oddballs, for example), but finally moved on to explosive disparity.

By 1978, the new conception had settled, as symbolized by Whittington's interpretation of *Aysheaia*. The period thereafter, and continuing today—Act 5 of my drama—possesses a new calm, in shared confidence about the general status of the Burgess fauna. Yet this final act is no anticlimax in its unaltered conceptual scheme. For confidence has a great practical virtue—you can go forward on specifics without continual worry about basic principles. Hence, Act 5 has witnessed an extraordinary productivity in the resolution of Burgess organisms. Old mysteries have fallen like ranks of tin soldiers—not quite so easily as child's play (to continue the simile), but with much greater efficiency now that a firm framework guides a coherent effort. The reconstructions of the last decade include some of the strangest and most exciting of Burgess creatures. I can hardly wait to read Act 6.

THE ONGOING SAGA OF BURGESS ARTHROPODS

Orphans and Specialists

At the end of 1978, the scorecard for soft-bodied arthropods spoke strongly for uniqueness and disparity. Four genera—*Marrella, Yohoia, Burgessia,* and *Branchiocaris*—had been orphaned within the arthropods. Only *Canadaspis* (and perhaps *Perspicaris*) belonged to a modern group; *Naraoia* had been reclassified as a trilobite, but as a surpassingly odd member of the group, and the prototype of a new order. *Opabinia* had been tossed out of the arthropods altogether, and *Aysheaia* lay in limbo. A good beginning, but not yet imbued with the convincing weight of numbers. As I argued above, the "big" questions of natural history are answered as relative frequencies. More data were required—something approaching a complete compendium of Burgess arthropods. Act 5 has now fulfilled this need, and the revisionary pattern has held, in spades.

In 1981, Derek Briggs continued his dispersion of the bivalved arthropods into a series of orphaned groups (with *Canadaspis* holding increasingly lonely vigil as a true crustacean). Briggs used all twenty-nine specimens to decide the fate of *Odaraia,* the largest bivalved arthropod in the Burgess Shale (up to six inches long). At the front of its head, and extending beyond the carapace, *Odaraia* bears the largest eyes of any Burgess arthropod (figure 3.44). Yet Briggs could find only one other structure on the head—a single pair of short ventral appendages behind the mouth. (This arrangement, with no antennae and only one post-oral pair of appendages, is unique, and would be sufficient in itself to mark *Odaraia* as an orphan among arthropods. But the head is not well preserved under the

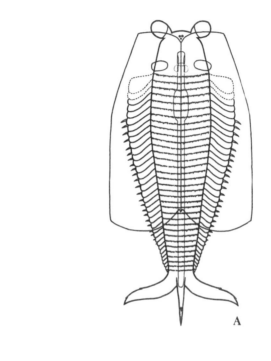

A

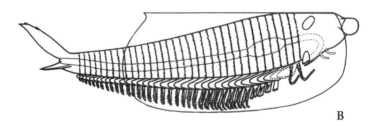

B

3.44. Reconstruction of the arthropod *Odaraia* by Briggs (1981a). (A) Top view, showing the bivalved carapace as transparent so that the soft anatomy may be revealed beneath. Note the projection of the eyes in front of the carapace, and the arrangement of the three-pronged tail behind. (B) Side view.

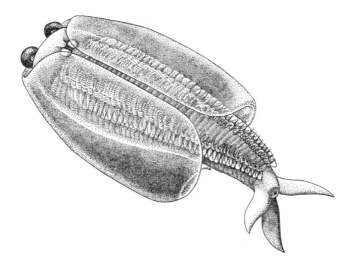

3.45. *Odaraia*, swimming on its back. The numerous biramous appendages can be seen through the transparent tubular carapace. Note also the large eyes in front, the curious three-pronged tail behind, and the single pair of feeding appendages behind the mouth. Drawn by Marianne Collins.

strong carapace of *Odaraia*, and Briggs was not confident that he had been able to resolve all structures.) The trunk, enclosed by the large carapace for more than two-thirds of its length, contained up to forty-five limb-bearing segments. The limbs, except perhaps for the first two pairs, are typically biramous.

Odaraia also exhibits two unique and peculiar specializations. This animal bears a three-pronged tail (figure 3.45), with two lateral flukes and one dorsal projection—a bizarre structure that evokes images of sharks or whales, rather than lobsters. Nothing similar exists in any other arthropod. Second, the bivalved carapace is not flattened, but essentially tubular. Moreover, Briggs argued that the relatively short appendages did not extend beyond the tube—and furthermore, that the two valves forming the tube probably couldn't gape widely enough to let the appendages protrude from any ventral opening. Clearly, *Odaraia* did not walk on the sea floor. Briggs wrote: "The combination of an essentially tubular carapace and a telson bearing these large flukes is unique among the arthropods" (1981a, p. 542).

Briggs performed a functional study and united these two peculiarities to infer a mode of life for *Odaraia*. He argued that *Odaraia* swam on its back, using its three-pronged tail for stabilization and steering, and its carapace as a filtering chamber for capturing food. Water could be taken in at one end; the appendages would extract food particles and pass the depleted stream out the other end of the carapace.

Briggs had proven once again that the watchword for Burgess arthropods was "uniquely specialized," not "primitively simple." In September 1988, Derek wrote to me, in an assessment of his 1981 monograph: "*Odaraia* turned out to be not only taxonomically unusual but, more importantly in my view, *functionally unique* among the arthropods."

Also in 1981, David Bruton published his monograph on *Sidneyia*, already discussed on pages 87–96. The resolution of *Sidneyia* set an important milestone in the study of Burgess arthropods for two reasons. First, *Sidneyia* had long acted as a focus or symbol for the fauna. Walcott regarded this genus as the largest of Burgess arthropods (we now know that the soft-bodied trilobite *Tegopelte* and one or two of the bivalved arthropods were bigger). Moreover, he mistakenly assumed that a spine-studded appendage, found separately, fitted onto the head of *Sidneyia* (for he knew nothing else big enough to carry such an appendage). With this addition *Sidneyia* was not only large, but also fierce. Since our culture values these traits, *Sidneyia* attracted attention. (A psychologist friend of mine explains our society's fascination with dinosaurs by a simple list—"big, fierce, and extinct." *Sidneyia*, in Walcott's reconstruction, is all three). In Bruton's revision, *Sidneyia* is still a predator, but the pair of limbs belongs to *Anomalocaris. Sidneyia* carries no feeding structures on its head.

Second, *Sidneyia* was the first form to be redescribed in the final, potentially coherent group of Burgess arthropods—the so-called "merostomoids." Hope had surely faltered for placing any major Burgess assemblage in a modern group, but the "merostomoids" represented a last gasp and opportunity for traditionalism. Merostomes are a group of marine arthropods including modern horseshoe crabs and fossil eurypterids. They are united with spiders, scorpions, and mites into one of the four great arthropod groups, the Chelicerata. The basic merostome body plan—more clearly shown by eurypterids, than by horseshoe crabs—includes a strong head shield, a trunk of several broad segments equal in width to the head, and a narrower tail, often forming a spike. Several Burgess genera, including *Sidneyia,* share this basic form.

Bruton dashed the final hope for traditionalism by showing that *Sidneyia* could not be a close relative or ancestor of merostomes. The "merostomoid" body did not define a coherent evolutionary group, but a series of disparate creatures united only by what our jargon calls a symplesiomorphic (or "shared primitive") trait. Shared primitive traits are ancestral for large groups, and therefore cannot define subgroups within the entire assemblage. For example, rats, people, and ancestral horses do not form a genealogical group within the mammals just because all have five toes. Five toes is an ancestral trait for Mammalia as a whole. Some creatures retain this initial condition; many others evolve modifications. The "merostomoid" body form is a shared primitive trait of many arthropods. True genealogical groups, by contrast, are based on shared *derived* characters—the unique specializations of their common ancestors.

True chelicerates have six pairs of appendages, and no antennae, on their head shield. *Sidneyia* could not be more different in this crucial respect. Its head (figure 3.46) bears one pair of antennae, and no other appendages! Bruton came to regard *Sidneyia* as a curious mosaic of characters. The first four of nine body segments carry uniramous walking legs like those of merostomes. But the five posterior segments bear ordinary biramous appendages, with gill branches and walking legs. The "tail" piece, formed of three cylindrical segments and a caudal fan, looks more crusta-

3.46. Two views of *Sidneyia:* top, as seen from below, showing the form of the limbs and the attachment of eyes and antennae; and bottom, as seen from above. Drawn by Marianne Collins.

cean than merostomoid. Bruton found ostracodes, hyolithids, and small trilobites in *Sidneyia*'s gut, and interpreted the animal as a bottom-dwelling carnivore. But with no feeding appendages on the head, and a strong, tooth-lined food groove between the legs, *Sidneyia* presumably fed like most arthropods, by passing food toward the mouth from the rear, not by searching and grasping from the front.

The year 1981 was pivotal for Burgess arthropods, and for the final dispersal of the last remaining "merostomoid" hope. For, in the same year of *Odaraia* and *Sidneyia*, Whittington published his "mop-up" monograph, "Rare Arthropods from the Burgess Shale, Middle Cambrian, British Columbia." Most or all of these animals had fallen (or would have fitted, had they been known at the time) into the "merostomoids." But Whittington could reconstruct not one as a chelicerate. All became orphans, unique arthropods unto themselves.

Molaria has a deep head shield, shaped like a quarter sphere, followed by eight trunk segments diminishing in size toward the rear, and capped by a cylindrical telson with a very long, jointed posterior spine, extending back

3.47. *Molaria*, a unique arthropod of "merostomoid" form (Whittington, 1981).

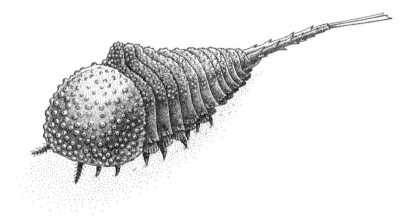

3.48. The tuberculate arthropod *Habelia*. Drawn by Marianne Collins.

more than the length of the body (figure 3.47). This basic form is faultlessly "merostomoid," but the head bears a pair of short antennae, followed by three pairs of biramous appendages.

Habelia has the same basic shape as *Molaria*, but Whittington also described an impressive set of differences, some of high taxonomic significance. The carapace is covered with tubercles—a superficial though visually striking difference (figure 3.48). The trunk has twelve segments, with no cylindrical telson. The extended tail spike, ornamented with barbs and ridges, is unsegmented, but has a single joint about two-thirds of the way back. The head has a pair of antennae and only two pairs of subsequent ventral appendages. The first six trunk segments bear biramous appendages, but the last six probably bore gill branches only (in *Molaria* all eight body segments bear biramous appendages.)

Whittington also discovered a new arthropod genus—a complex, tiny creature less than a half inch in length (figure 3.49). This unique and peculiar animal, named *Sarotrocercus,* has a head shield followed by nine body segments and a tail spike with a tuft of spines at its tip. A large pair of eyes, borne on stalks, protrudes from the bottom front end of the head shield (*Molaria* and *Habelia* are blind). In addition, the head carries one pair of thick, strong appendages terminating in a two-pronged segment. Whittington also found ten very different pairs of appendages (one pair on the head and one on each of the nine body segments)—long comblike structures, presumably gill branches, but without any evident trace of a leg branch. Whittington reconstructed *Sarotrocercus* as a pelagic animal,

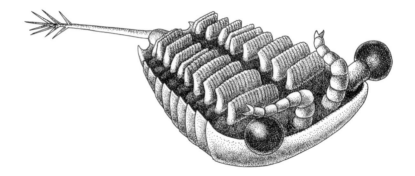

3.49. The tiny arthropod *Sarotrocercus,* swimming on its back. Note the large eyes, the strong pair of feeding appendages, and the gill branches, presumably used for swimming, on the body segments behind. Drawn by Marianne Collins.

swimming on its back with *Amiskwia* and *Odontogriphus* among the rare Burgess organisms that probably lived in the water column above the stagnant basin that received the mudslide.

Actaeus, based on a single specimen two inches long, has a head shield with a marginal eye lobe, followed by eleven body segments and an elongate, triangular terminal plate (figure 3.50). The head bears a pair of remarkable appendages, each with a stout initial portion, bent and extending downward, ending in a group of four spines. Two very long whiplike extensions attach to the inner border of the last segment, and run down and back. Behind this structure, the head probably carried three pairs of ordinary biramous appendages.

Alalcomenaeus has a basically similar look and arrangement of appendages (see figure 3.50), and may be related to *Actaeus.* A head shield, bearing a marginal eye lobe, is followed by twelve body segments and an ovate terminal plate. The head bears a pair of large appendages, each with a broad initial section followed by a long thin extension—not nearly so complex as in *Actaeus,* but similar in style and position. The head also carries three pairs of biramous appendages. One specimen reveals an impressive set of spines on the inner surfaces of the walking legs—in proper position for passing food forward to the mouth. "These remarkable appendages," Whittington wrote, "suggest a benthic scavenger, able to hold on to, and tear up, a carcass" (1981a, p. 331).

Aside from a very tentative relationship between *Actaeus* and *Alalcomenaeus*, each of the five genera presented a highly specialized design based on unique features and arrangements of parts. Whittington concluded, echoing the now-familiar Burgess story:

> Many new and unexpected features have been revealed, and the morphological gaps between species greatly enlarged. Each, with rare exceptions, shows a most distinctive combination of characters. The selection [of genera] dealt with here adds further to the range of morphological characters in the non-trilobite arthropods, and to the variety of distinctive combinations of characters (1981a, p. 331).

In 1983, Bruton and Whittington combined to deliver the *coup de grâce* by describing the last two major Burgess arthropods—the large *Emeraldella* and *Leanchoilia*, last two members of Størmer's discredited Merostomoidea.

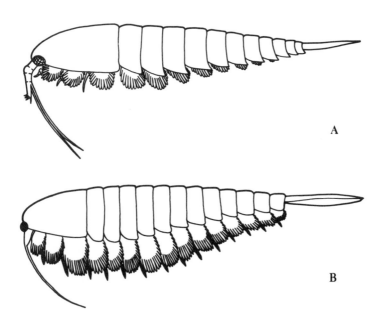

3.50.   Two arthropods that may be closely related (Whittington, 1981). (A) *Actaeus*. (B) *Alalcomenaeus*.

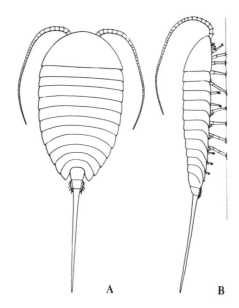

3.51. *Emeraldella*, seen from above (A), and from the side (B), resting on the bottom. The very small gill branches of the biramous appendages indicate that this animal walked on the sea floor.

Emeraldella possesses the basic "merostomoid" form, but accompanied by yet another set of unique structures and arrangements. The typical head shield bears a pair of very long antennae, curving up and back, followed by five pairs of appendages, the first short and uniramous, the last four biramous (figure 3.51). The first eleven trunk segments are broad, though progressively narrowing toward the rear, and each bears a pair of biramous appendages. The last two segments are cylindrical, and a long unjointed tail spine extends at the rear.

Leanchoilia also shares the superficiality of general "merostomoid" shape, with a triangular head shield (terminating in a curious, upturned "snout"), followed by eleven body segments, narrowing and curving backward beyond the fifth. A short triangular tail spine with lateral spikes caps the nether end (figure 3.52). *Leanchoilia* bears thirteen pairs of biramous appendages, two at the rear of the head shield, one on each of the eleven body segments.

But *Leanchoilia* also possesses the most curious and interesting appendage of any Burgess arthropod—an exaggerated version of the frontal structure of *Actaeus,* a possible relative. Borrowing a term from *Yohoia,* and in the absence of any appropriate technical name, Bruton and Whittington simply called this structure the "great appendage." Its basal part contains four stout segments facing down at first, but bending through ninety degrees to run forward. The second and third segments end in very long, whiplike extensions, annulated over the last half of their length. The fourth segment has a tapering shaft ending dorsally in a group of three claws, and extending ventrally as a third whiplike structure with annulations. The different orientations of various specimens indicate that this great appendage was hinged at its base (figure 3.53) and could extend forward, to help *Leanchoilia* repose on the substrate (figure 3.54), or bend back, perhaps to reduce resistance in swimming. Further evidence for swimming as a primary mode of life comes from the biramous appendages. Unlike *Emeraldella,* with its long walking legs and small gill branches, *Leanchoilia* bears such large gill branches that they form a veritable curtain of overlapping, lamellate lobes, completely covering and extending beyond the shorter leg branches underneath.

The completed redescription of all "merostomoid" genera prompted Bruton and Whittington to reflect upon the incredible disparity uncovered

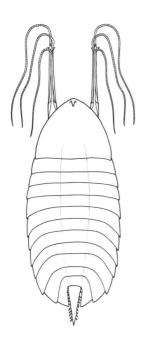

3.52. Top view of *Leanchoilia.* Note the three whiplike extensions of the great appendage in front and the triangular tail spine behind.

beneath a superficial similarity of outward form. Consider only the arrangement of appendages on the head—an indication of original patterns in segmentation, and a guide to the deep anatomical structure of arthropods. *Sidneyia* has a pair of antennae and no other appendages. *Emeraldella* also bears pre-oral antennae, but has *five* additional pairs of appendages behind the mouth, one uniramous and four biramous. *Leanchoilia* does not possess antennae, but bears its remarkable "great appendages," followed by two biramous pairs behind the mouth.

The Burgess had been an amazing time of experimentation, an era of such evolutionary flexibility, such potential for juggling and recruitment of

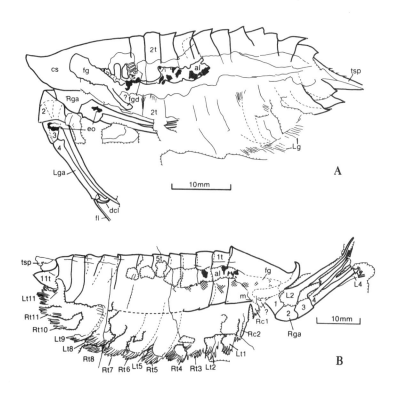

3.53. Camera lucida drawings of two specimens of *Leanchoilia*. The great appendages are labeled *Lga* and *Rga,* and their major segments are numbered. (A) The great appendages are folded back, presumably in the swimming position; the right appendage is flat against the body, with the left just below. A trace of the gut, or alimentary canal *(al)* and the tail spine *(tsp)* are visible. (B) The appendages extend forward, in the feeding position.

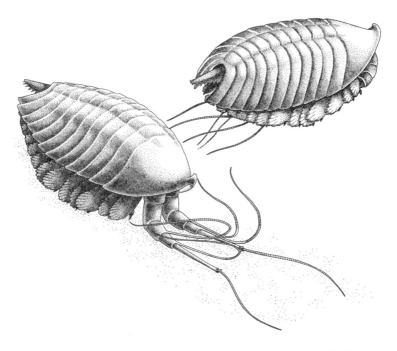

3.54. Two views of *Leanchoilia:* top, in swimming position, with the great appendages folded back and the whiplike tentacles extending beyond the length of the body; and bottom, with the great appendages extending forward to aid the animal in resting on the bottom. Drawn by Marianne Collins.

characters from the arthropod grabbag, that almost any potential arrangement might be essayed (and assayed). We now recognize clear groups, separated by great morphological gulfs, only because the majority of these experiments are no longer with us. "It was only later that certain of these solutions were fixed in combinations that allow the present arthropod groups to be recognized" (Bruton and Whittington, 1983, p. 577).

A Present from Santa Claws

Bureaucratic entanglement provides one possible benefit amidst its own distinctive and inimitable brand of frustration. You sometimes get so angry that you do something useful as an end run around intransigence. As the old motto goes, Don't get mad, get even. When Des Collins, after sublime patience and deep entanglement, was denied permission to excavate at

Walcott's quarry and allowed only to gather specimens from the talus slope (under further restrictions and nearly endless delays), he realized that he would have to shift his Burgess interests elsewhere.*

Collins therefore began to search for Burgess equivalents in surrounding areas, where collection and excavation might be permitted. He succeeded abundantly, finding soft-bodied fossils at more than a dozen additional nearby localities. Most of these assemblages contain the same species as Walcott's quarry, but Collins made a few outstanding discoveries of his own. At a locality five miles south of Walcott's quarry (Collins, 1985), and one hundred feet below in stratigraphic sequence, Collins made the find of the decade—a large arthropod with so many spiny appendages on its head that Collins, following an old tradition of field work, gave it a nickname. As Walcott had called *Marrella* the "lace crab," Collins dubbed his discovery "Santa Claws." Working with Derek Briggs, Collins has now formalized and honored this name in his technical description (Briggs and Collins, 1988). "Santa Claws" is now, officially, *Sanctacaris,* which means almost the same thing.

Sanctacaris has a bulbous head shield, wider than long and extending laterally as a flat, triangular projection on each side (figure 3.55). The body bears eleven broad segments, the first ten with a pair of biramous appendages. A wide, flat telson caps the rear end. The combination of large lamellate gill branches on the body appendages and a broad telson well designed for stabilization and steering indicates that *Sanctacaris* probably favored swimming over walking.

The striking suite of head appendages identifies this relatively large Burgess arthropod (up to four inches long) as a carnivore specialized for direct pursuit. The first five pairs make a coordinated and formidable array that inspired Collins's field name. They are biramous, with the outer branches reduced to antenna-like projections (not gills) and the inner branches arranged as a fierce-looking set of jointed feeding appendages with sharp spines on the inner borders. These feeding branches gain in length from front to back, starting with four segments on the first pair, and increasing to eight or more on the fifth. The sixth pair, different in both form and position, lies behind the first five and well to the side. The outer

---

*I am as committed as anyone to "ecology" (in the vernacular and political meaning of leaving nature alone), and I certainly believe in respecting the nearly sacred integrity of national parks. But a fossil on the ground is worth absolutely nothing. It is not an object of only pristine beauty, or a permanent part of any natural setting (especially for fossils exposed in quarry walls). If free on the ground, it will probably be cracked and frost-heaved into oblivion by the next field season. Controlled collecting and scientific study are the proper roles, intellectually and ethically, for the Burgess fossils.

branch is, again, similar to an antenna in form, but much larger than the corresponding branch of the five feeding appendages. The inner branch is short, but terminates in an impressive fringe of radiating spines.

One might think at first assessment, Oh, just another of those Burgess "merostomoids"—with a forest of head appendages as its distinctive specialization, just as *Habelia* has its tubercles, *Sidneyia* its stout walking legs, and *Leanchoilia* its great appendage. Interesting, but not my advertised "find of the decade."

Not so. The difference between *Sanctacaris* and the others is taxonomic, and conceptually stunning: *Sanctacaris* seems to be a genuine chelicerate, the first known member of a line that eventually yielded horseshoe crabs, spiders, scorpions, and mites. *Sanctacaris* bears the requisite six pairs of appendages on its head. None of these appendages has been specialized to form the distinctive claw, the chelicera, that defines and names the group, but the absence of a structure early in the geological run of a group may simply mean that such a specialization has not yet evolved.

Briggs and Collins (1988) have also identified other derived chelicerate characters (including the differentiation of head from body appendages, and the position of the anus), thus corroborating the status of *Sanctacaris* by more than a single feature. They state:

> Such a combination is unique to the chelicerates. The apparent lack of chelicerae, an advanced character present in all other chelicerates, is consistent with the primitive biramous appendages on both the head and trunk. It places *Sanctacaris* in a primitive sister group to all other chelicerates.

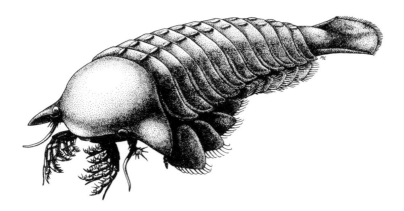

3.55. *Sanctacaris.* Drawn by Marianne Collins.

The limbs of modern chelicerates are uniramous, with the outer branch lost on the head appendages (yes, the walking legs of spiders are all on the prosoma, or head portion), and the inner branch lost on the trunk (yes again, spider gills are on the opisthosoma, or body portion). *Sanctacaris*, by preserving the full set of possibilities before selective elimination in later specialized lines, serves as an interesting structural precursor for its great group.

But the chief excitement of *Sanctacaris* lies in its key role in completing the fundamental argument for Burgess arthropods. With the discovery of *Sanctacaris*, we now have, in the Burgess, members of *all four* great arthropod groups—trilobites in fair abundance, crustaceans represented by *Canadaspis*, uniramians by *Aysheaia** (accepting Robison's interpretation, as I do), and chelicerates by *Sanctacaris*. They are all there—but so are at least thirteen other lineages (and perhaps as many again yet to be described) of equal morphological uniqueness. Some of these thirteen are among the most specialized *(Leanchoilia)* or, at least by numbers, the most successful *(Marrella)* of Burgess arthropods. I challenge any paleontologist to argue that he could have gone back to the Burgess seas and, without the benefit of hindsight, picked out *Naraoia, Canadaspis, Aysheaia,* and *Sanctacaris* for success, while identifying *Marrella, Odaraia, Sidneyia,* and *Leanchoilia* as ripe for the grim reaper. Wind back the tape of life, and let it play again. Would the replay ever yield anything like the history that we know?

CONTINUING THE MARCH OF WEIRD WONDERS

The last decade, so satisfying for arthropods, has also witnessed the resolution of two additional weird wonders—unique and independent anatomies that would merit classification as separate phyla if we felt comfortable about bestowing so high a taxonomic rank on a single species (see Briggs and Conway Morris, 1986, for a list of such Burgess creatures still unstudied). These two works may be the most elegant and persuasive in the entire

---

*The status of the Onychophora, probable taxonomic home of *Aysheaia*, remains controversial. Some experts regard the Onychophora as an entirely separate phylum, no closer to the uniramians than to any other group of arthropods. If this solution is correct, my argument here is wrong. The two other major solutions both support my argument: first, that Onychophora should rank within the Arthropoda on the uniramian line; second (and probably the predominant view), that onychophorans deserve separate status, but lie closer to the uniramians than to any other group of arthropods. (This last argument assumes a separate evolutionary origin for several, perhaps all four, of the great arthropod lines—with uniramians arising in genealogical proximity to onychophorans.)

Burgess canon. They stand as a fitting end to my play, for they combine the greatest intellectual and aesthetic satisfaction with an assurance that this particular drama has no foreseeable end.

Wiwaxia

When I asked Simon Conway Morris why he had chosen to work for many years on so complex a beast as *Wiwaxia*, he replied, with welcome frankness, that Harry and Derek had both done their "blockbusters," and he wanted to prove that he could also write a "strict monograph in the tradition of the others." (I regard this statement as overly modest. Simon's 1977 and 1979 works on priapulids and polychaetes are true and extensive monographs. But each treats several genera, and therefore cannot give the exhaustive treatment to any one species that Whittington provided for *Marrella splendens*, or Briggs for *Canadaspis perfecta*.) Perhaps Simon felt unfulfilled in choosing such rare creatures for his first run through the weird wonders that he could write only short, separate papers on five examples. In any case, his monograph on *Wiwaxia* is a thing of beauty, and the original source of my interest in writing about the Burgess Shale (Gould, 1985b)—for which, Simon, my greatest thanks once again.

Wiwaxia is a small creature, shaped as a flattened oval (a well-rounded pebble in a stream comes to mind), about an inch long, on average, with a two-inch maximum. The simple body is covered with plates and spines called sclerites—except for the naked ventral surface that rested on the substrate as *Wiwaxia* crawled across the sea floor. Walcott had shoehorned *Wiwaxia* into the polychaete worms, mistaking these sclerites for superficially similar structures in a well-known marine worm, whose technical and common names convey such different impressions—*Aphrodita*, the sea mouse. But *Wiwaxia* has no body segmentation and no true setae (the hairlike projections of polychaetes)—and therefore lacks both defining traits of the group. Like so many Burgess animals, *Wiwaxia* is an anatomy unto itself. *Wiwaxia* is also inordinately difficult to reconstruct, because the sclerites spread over the rock surface in a horribly confused jumble as the fossil compressed on its bedding plane. In figure 3.56, a camera lucida drawing of the most coherent specimen in the most convenient orientation provides a good idea of the problems involved. Simon's resolution of *Wiwaxia* is one of the great technical achievements of the Burgess research program.

The sclerites of *Wiwaxia*, the key to this reconstruction, grew in two different styles; flattened scales, ornamented with parallel ridges, cover most of the body, while two rows of spines emerge from the top surface,

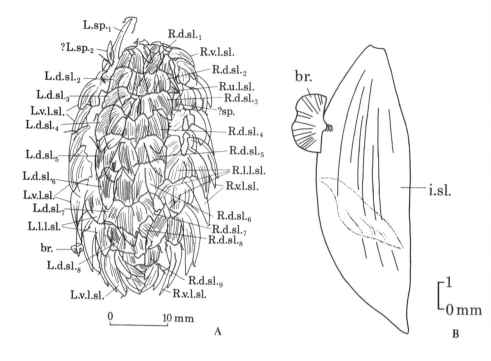

3.56. (A) Camera lucida drawing of a complete specimen of *Wiwaxia*. Note the complex intermingling of the compressed sclerites. The labels, which need not concern readers here, identify individual sclerites. For example, *R.d.sl.*₁ (top right) is a right, dorsal sclerite *(sl.)* of the first row. *L.sp.*₁ (top left) is the first spine on the left side. (B) Enlargement of one particularly interesting sclerite (located in A at the lower left, next to the label *br.*). A small brachiopod *(br.)* affixed itself to the sclerite during the life of this *Wiwaxia* specimen. Using such evidence, we can reconstruct the life style of this animal. It could not have lived by burrowing under the substrate, for such a habit would have killed the brachiopod.

one on each side of the central axis (figures 3.57 and 3.58). The scales display a symmetrical and well-ordered tripartite pattern: (1) a field of overlapping plates, on the top surface, arrayed as six to eight parallel rows (figure 3.57A); (2) two regions on each side (figure 3.57B), with two rows of plates pointing upward and two rows pointing backward; (3) a single bottom row of crescent-shaped sclerites forming a border between the ornamented upper body and the naked belly.

The two rows of seven to eleven elongate spines arise from the upper row of sclerites on each side, near the border with the plates of the top surface. The spines project upward and presumably acted as protection against predators, as indicated by their breakage in several specimens (during the animal's life, not after burial).

Simon could see little of *Wiwaxia*'s internal anatomy beyond a straight gut near the ventral surface—further evidence, combined with the naked belly and spines pointing upward, for the animal's orientation in life. But one internal feature may be crucial both for understanding *Wiwaxia* and for a general interpretation of the Burgess fauna. About five millimeters from the front end, Conway Morris found two arc-shaped bars, each carry-

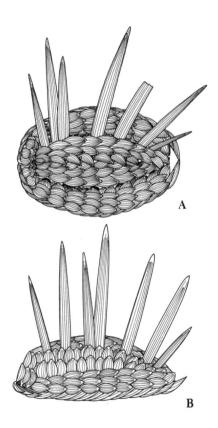

3.57. Reconstructions of *Wiwaxia* by Conway Morris (1985). (A) Top view: one of the two rows of spines has been omitted (note the blackened areas of their insertion) so that the sclerites can be seen better. (B) Side view: the front end is at the left.

ing a row of simple, conical teeth directed toward the rear (figure 3.59). The front bar bears a notch at its center, marking a toothless area between the side regions, each with seven or eight teeth. The rear bar has a more curved but smoother front margin, and teeth all along the back edge. These structures were probably attached to the bottom of the gut. In view of their form and their position near the animal's front end, their interpretation as feeding devices—"jaws," if you will—seems secure.

In attempting to gather and integrate all the evidence, Conway Morris proceeded as far as possible beyond the basic anatomy of *Wiwaxia*, probing for hints wherever he could extract some precious information—from growth, from injury, from ecology, from preservation. Small specimens either carry relatively small spines or lack them entirely—thus providing a rare Burgess example of change in form with growth. Two juxtaposed specimens seem to represent an act of molting by one individual, not two animals accidentally superimposed by the Burgess mudslide: the smaller specimen is shrunken and elongate, as if the large body had just crawled out, leaving its old skin behind as "a vacated husk." Small brachiopod shells, occasionally found attached to a sclerite, indicate that *Wiwaxia* crawled along the top of the sediment, and did not burrow underneath, where the permanent hitchhikers could not have survived. Patterns of breakage in spines point to the activity of predators and to the possibility of

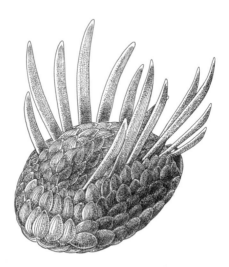

3.58. *Wiwaxia* as it might have crawled on the sea floor. Drawn by Marianne Collins.

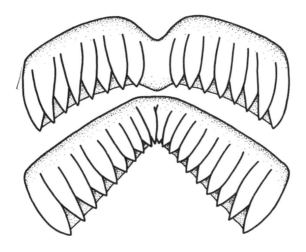

3.59.   The jaw apparatus of *Wiwaxia* (Conway Morris, 1985).

escape. Small spines occasionally found in an otherwise large and uniform row indicate the possibility of regeneration after breakage, or of orderly patterns in replacement (as in the shedding and cycling of teeth in vertebrates without a permanent dentition). The presence of "jaws" suggests a life spent scraping algae or gathering detritus on the substrate.

Put all these bits and pieces together, and *Wiwaxia* emerges as a complete, working organism—a herbivore or omnivore, living on small items of food collected from the sediment surface as it crawled along the sea floor.

But if all these guides had enabled Conway Morris to reconstruct *Wiwaxia*'s mode of life, he could find no similarly persuasive clues to homology, or genealogical relationship with any other group of organisms. With no setae or appendages and no segmentation, *Wiwaxia* is neither an arthropod nor an annelid. The jaw displays an intriguing similarity to the feeding apparatus of mollusks, called a radula, but nothing else about *Wiwaxia* even vaguely resembles a clam, snail, octopus, or any other mollusk living or dead.* *Wiwaxia* is another Burgess oddball, perhaps closer to the Mollusca than to any other modern phylum, if its jaw can be homologized with the molluscan radula—but probably not very close.

---

*A small and little-known molluscan group called the Aplacophora does seem more similar in its elongate, wormlike body, sometimes covered with plates or spicules, but Conway Morris enumerates an impressive list of detailed differences in his monograph.

Anomalocaris

I could not have made up a better story to illustrate the power and extent of the Burgess revision than the actual chronicle of *Anomalocaris*—a tale of humor, error, struggle, frustration, and more error, culminating in an extraordinary resolution that brought together bits and pieces of three "phyla" in a single reconstructed creature, the largest and fiercest of Cambrian organisms.

The name *Anomalocaris,* or "odd shrimp," predates the discovery of the Burgess Shale, for this is one of the few soft-bodied Burgess creatures endowed with parts solid enough for preservation in ordinary faunas (the spicules of *Wiwaxia* are another example). The first *Anomalocaris* were found in 1886 at the famous *Ogygopsis* trilobite beds, exposed on the next mountain over from the Burgess Shale. In 1892, the great Canadian paleontologist J. F. Whiteaves described *Anomalocaris* in the *Canadian Record of Science* as the headless body of a shrimplike arthropod. Walcott accepted the standard view that this fossil represented the rear end of a crustacean, with the long axis as the trunk and the ventral spines as appendages (figure 3.60). Charles R. Knight followed this tradition in his famous painting of the Burgess fauna (see figure 1.1), where he constructed a composite organism by attaching *Anomalocaris* to *Tuzoia,* one of the bivalved arthropod carapaces that lacked associated soft parts and was therefore a good candidate for the cover of *Anomalocaris*'s unknown head.

But this official name-bearer of *Anomalocaris* provides only one piece of

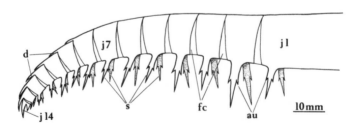

3.60. The fragment of a segmented creature originally called *Anomalocaris* in 1886 (Briggs, 1979). For many years this fossil was considered to represent the trunk and tail of an arthropod. It has now been correctly identified as one of a pair of feeding appendages from the largest of all Cambrian animals.

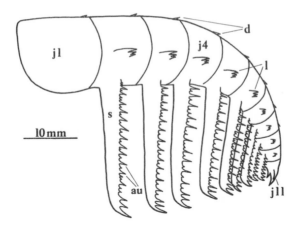

3.61. Reconstruction of appendage F by Briggs (1979). Walcott originally described this structure as a feeding limb of *Sidneyia*. Briggs reinterpreted it as an appendage of a giant arthropod. Recent research shows that appendage F is actually one of a pair of feeding organs from the largest known Cambrian animal.

our story. Three other structures, all named by Walcott, play central roles in this complex tale.

1. The head of *Sidneyia*, the arthropod that Walcott named for his son Sidney and then described first among Burgess creatures (1911a), bears a pair of antennae and no other appendages. Walcott also found a large isolated arthropod feeding limb, later (1979) called "appendage F" (for feeding) by Derek Briggs (figure 3.61). *Sidneyia* was, in Walcott's judgment, the only Burgess creature large enough to carry such an appendage; its rapacious character also fitted well with Walcott's concept of *Sidneyia* as a fierce carnivore. So Walcott made the marriage without direct evidence, and joined appendage F to the head of *Sidneyia*. Bruton (1981) later determined that *Sidneyia*'s head shield does not contain enough space to accommodate such a structure.

2. Walcott's second paper (1911b), on the supposed jellyfish and holothurians (sea cucumbers of the echinoderm phylum) from the Burgess Shale, does not rank among his more accurate efforts. He described five genera. *Mackenzia* is probably a sea anemone and therefore a coelenterate in the same phylum as jellyfish, but Walcott placed this genus in his other group, the holothurians. A second creature turned out to be a priapulid worm (Conway Morris, 1977d). A third, *Eldonia*, still ranks as a peculiar

floating holothurian in the latest reconstruction (Durham, 1974), but I'll wager a reasonable sum that it will finally end up as another Burgess oddball.

Walcott named a fourth genus *Laggania,* and identified this fossil as a holothurian, on the basis of one specimen. He noted a mouth, and thought that it might be surrounded by a ring of plates. Poor preservation had effaced all the distinctive features of holothurians. Walcott admitted: "The body of the animal is so completely flattened that the tube feet are obscured, the outline of the ventral sole lost, and the concentric bands almost obliterated" (1911b, p. 52).

3. As a fifth and last genus, Walcott named the only Burgess jellyfish *Peytoia.* He described this peculiar creature as a ring of thirty-two lobes around a central opening. This series of lobes could be divided into four quadrants, with a larger lobe at each of the four corners of the squared-off ring, and seven smaller lobes between the corners in each quadrant. Walcott noted two short points on each lobe, projecting inward toward the central hole. He interpreted these structures as "points of attachment of the parts about the mouth, or possibly oral arms" (1911b, p. 56). Except for radial symmetry, Walcott found no trace of the defining characters of a jellyfish—no tentacles or concentric muscle bands. *Peytoia,* looking more like a pineapple slice than a medusa, made an awfully odd jellyfish. No true member of the group has a hole in the center. Nonetheless, Walcott's interpretation prevailed. The best-known modern reconstruction of the Burgess fauna, published in *Scientific American* several years after Whittington and colleagues began their revisions (Conway Morris and Whittington, 1979), shows *Peytoia* as a kind of Frisbee *cum* flying saucer *cum* pineapple slice, entering the scene from the west (figure 3.62).

Now who ever dreamed about a connection between the rear end of a shrimp, the feeding appendage of *Sidneyia,* a squashed sea cucumber, and a jellyfish with a hole in the center? Of course, no one did. The amalgamation of these four objects into *Anomalocaris* came as an entirely unanticipated shock. Moreover, the successful resolution did not emerge from this unimproved initial chaos. Several intermediate efforts, all basically erroneous but each supplying an important link in a developing story, preceded the successful conclusion.

Anomalocaris has been the nemesis of recent Burgess research. This creature eventually yielded its secret, but not until both Simon Conway Morris and Derek Briggs had committed their biggest mistakes in coping with its various parts. One cannot hope to do anything significant or original in science unless one accepts the inevitability of substantial error along the way. Three steps, however, did inch matters forward toward a resolu-

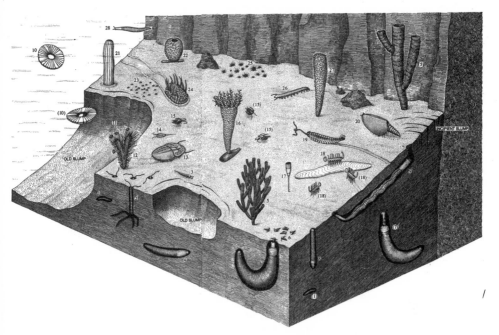

3.62. The best-known reconstruction of the Burgess Shale, drawn for the 1979 *Scientific American* article by Conway Morris and Whittington. Note priapulid worms in their burrows, and several Burgess oddballs—including *Dinomischus* (17), *Hallucigenia* (18), *Opabinia* (19), and *Wiwaxia* (24). In a major error, two jellyfish (10) are shown swimming in like pineapple slices from the west. This structure is actually the mouth of *Anomalocaris*. (From "The Animals of the Burgess Shale," by Simon Conway Morris and H. B. Whittington. Copyright © 1979 by Scientific American, Inc. All rights reserved.)

tion, whatever the longer lateral errors.

1. In 1978, Conway Morris applied Whittington's new techniques for distinguishing three-dimensional structure to *Laggania*, now regarded as a sponge rather than a holothurian. He took a dental microdrill to the counterpart of the unique specimen, and uncovered a pineapple slice of *Peytoia*, where Walcott had identified the indistinct mouth. Conway Morris stood on the threshold of the proper interpretation, but he guessed wrong. He considered the possibility that the "sponge" called *Laggania* was not a distinct creature, but a body attached to *Peytoia*, which would then become the centerpiece of a strange medusoid. But Conway Morris rejected this reconstruction because he regarded almost all Burgess organisms as discretely preserved, rather than disaggregated into parts. He wrote: "The vast majority of Burgess Shale fossils are preserved complete and it may reasonably be concluded that the body of *Laggania cambria* is not an

integral part of *Peytoia nathorsti,* but an extraneous addition to the medusoid which is interpreted here as a sponge" (1978, p. 130). He argued that
the association was simply an accident of deposition from the Burgess mudslide: "The association of the medusoid and sponge is presumably by chance.
The phyllopod bed was deposited as a series of turbidites, and it is likely
that after transport the two specimens settled together" (1978, p. 130).

Conway Morris guessed wrong about the reasons for a link between
Peytoia and *Laggania,* but he had uncovered (literally) a key association,
joining the first two of four pieces that would form *Anomalocaris.*

2. In 1982, Simon tried to grapple with the strangeness of *Peytoia* (Conway Morris and Robison, 1982). He called *Peytoia* "one of the most peculiar of Cambrian medusoids" (1982, p. 116), and even used the word
"enigmatic" in his title. Simon did not correctly resolve this beast, but he
cast doubt upon its affinity with medusoids, and thus kept the channels of
questioning wide open. Writing about the central hole, Conway Morris
and Robison concluded: "This feature is unknown in either living or fossil
cnidarians and may indicate that *Peytoia nathorsti* is not a cnidarian. Its
relationship with any other phylum would seem to be even more obscure"
(1982, p. 118).

3. *Anomalocaris* itself, Whiteaves's original rear end of a shrimp, had
been allocated to Derek Briggs in the original divvying up of the Burgess
Shale. It was, after all, supposed to be the body of an arthropod with a
bivalved carapace.

In 1979, Briggs published a provocative reconstruction of his assignment. He made two outstanding observations that contributed to the resolution of *Anomalocaris:*

First, he recognized that *Anomalocaris* was an appendage with paired
spines on its inner borders, not an entire body with appendages on its
ventral edges. If *Anomalocaris* was the trunk of an entire organism, then
some of the more than one hundred specimens should show traces of a gut,
and at least a few would be found with arthropod joints on their supposed
appendages.

Second, he argued that *Anomalocaris* and appendage F (Walcott's feeding limb of *Sidneyia*) were variants of the same basic structure, and probably belonged together. This conclusion, as we shall see, was not quite
correct, but Briggs's argument did properly unite two more pieces of the
Anomalocaris puzzle.

Beyond these important insights, Briggs's reconstruction was basically
erroneous, though spectacular. He continued to view both *Anomalocaris*
and appendage F as parts of an arthropod, conjecturing that *Anomalocaris*
was a walking leg, and appendage F a feeding structure, of a single giant

creature, probably more than three feet long! He called his paper "*Anomalocaris*, the Largest Known Cambrian Arthropod."

But Briggs was scarcely convinced by his own reconstruction. So many mysteries remained. He puzzled over the failure to find any sign, even fragmentary, of the giant body that supposedly held these appendages. Could a structure three feet long be entirely absent from a soft-bodied fauna? Briggs conjectured that such pieces might exist as organic sheets and films, thus far ignored for their lack of distinguishable structures. He wrote: "Large, previously unidentified, relatively featureless fragments of the body cuticle of *Anomalocaris canadensis* almost certainly await discovery on the scree slopes of Mt. Stephen" (1979, p. 657). Little did Derek realize that the body of *Anomalocaris* had been known and named since Walcott's time, but masquerading as the "holothurian" *Laggania*, later interpreted as a sponge with a jellyfish on top.

The Geological Survey of Canada expedition had discovered an odd specimen in the Raymond quarry, just above Walcott's phyllopod bed. Whittington had taken this large, ill-defined, and virtually featureless fossil and placed it in a drawer—hoping, I think, to bury it by the old cliché: Out of sight, out of mind. But he kept thinking about this peculiar fossil of a creature so much larger than anything else in the Burgess Shale. "I used to open the drawer and then close it," Harry explained to me. One day in 1981, he decided to excavate the fossil in the hope that some details of structure might be resolved. He dug into one end of the creature and, to his astonishment, found a specimen of *Anomalocaris* apparently attached and in place (figure 3.63). Harry told Derek Briggs about his discovery, and Derek simply couldn't believe it. The excavated object was surely *Anomalocaris*, but, like Simon's interpretation of the jellyfish *Peytoia* on the sponge *Laggania*, perhaps this specimen of *Anomalocaris* had been accidentally entangled with a large sheet of something else as the mudslide coalesced.

Soon afterward, Whittington and Briggs were studying a suite of specimens borrowed from the Walcott collections. These slabs showed relatively featureless blobs and sheets that had never attracted much attention, including the body of *Laggania* with *Peytoia* on top. On a single momentous day—the positive counterpart (in the vernacular, not technical, sense) of another key Burgess moment, nearly a decade before, when Whittington had cut through the head and sides of *Opabinia* and found nothing underneath—they excavated and found *both Peytoia* and appendage F as organs of a larger creature.

As they assimilated this greatest of all Burgess surprises, and kept finding *Peytoia* and appendage F in the same association on other slabs, Harry and

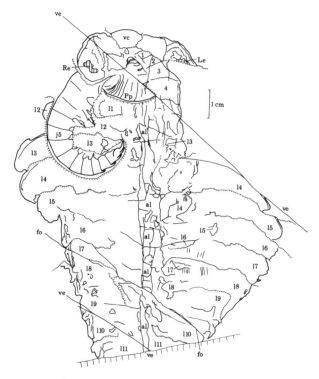

3.63. The specimen dissected by Harry Whittington that revealed the true nature of *Anomalocaris*. In this camera lucida drawing, the mouth misidentified by Walcott as the jellyfish *Peytoia* is at top center (labeled *Pp*); the oblique line *(ve)* just above it represents a crack in the rock. The structure originally named *Anomalocaris* is the curved feeding appendage just to the left of the mouth with its middle segment labeled *j5*. Also visible is the trace of the central gut, or alimentary canal *(al)*.

Derek realized that they had resolved a forest of problems into one creature. *Peytoia* was no jellyfish, but the mouth of the large beast, attached to the ventral surface near the front. Appendage F was not one member of a large sequence of repeated limbs on an arthropod; rather, two appendage F's formed a single pair of feeding organs attached, in front of the mouth, to the bottom end of the new animal.

But Whittington's specimen back in England bore *Anomalocaris*, not appendage F, in this frontal position (see figure 3.63). When he dissected this specimen more fully, he found traces of both the *Peytoia* mouth and a second *Anomalocaris*, forming a pair of feeding organs in the same posi-

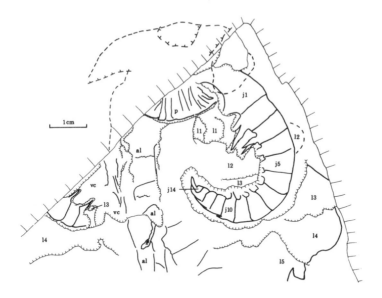

3.64. The key specimen of *Anomalocaris* further dissected to reveal parts of both feeding appendages. This is the other slab, and therefore a mirror image, of part of the specimen represented in figure 3.63. Note the mouth (labeled *p*) and the first discovered appendage *(j1–j14)*. But now a trace of the second feeding appendage has been excavated at the lower left, just below the oblique line representing the crack in the rock.

tion as the appendage-F pairs on the specimens in Washington (figure 3.64).

All the pieces had finally come together. From four anomalies—a crustacean without a head, a feeding appendage that didn't fit, a jellyfish with a hole in the middle, and a squashed sheet that had bounced from one phylum to another—Whittington and Briggs had reconstructed two separate species of the single genus *Anomalocaris. Laggania* was a squashed and distorted part of the body; *Peytoia,* the mouth surrounded by a circlet of toothed plates, not a series of lobes with hooks; *Anomalocaris* the pair of feeding organs in one species *(Anomalocaris canadensis);* appendage F a feeding organ in the second species (*Anomalocaris nathorsti,* borrowing the old trivial name of *Peytoia*). The uncompromising rules of nomenclature, honoring oldest first, required that the entire genus be called *Anomalocaris,* to recognize Whiteaves's original publication of 1892. But what a happy and appropriate imposition in this case—an "odd shrimp" indeed!

Since the organ originally named *Anomalocaris* can be up to seven inches in length when extended, the entire animal must have dwarfed nearly everything else in the Burgess Shale. Whittington and Briggs estimated the biggest specimens as nearly two feet in length, by far the largest of all Cambrian animals! A recent reconstruction of the whole fauna (Conway Morris and Whittington, 1985), basically an update of the 1979 *Scientific American* version, has replaced the pineapple-slice *Peytoia* that used to angle in from the west (see figure 3.62) with a large and menacing *Anomalocaris*, purposefully advancing from the east (figure 3.65).

Whittington and Briggs published their monograph on *Anomalocaris* in 1985, a fitting triumph to cap what may be the most distinguished and important series of monographs in twentieth-century paleontology. The long oval head of *Anomalocaris* bears, on the side and rear portion of its dorsal surface, a large pair of eyes on short stalks (figure 3.66). On the ventral surface, the pair of feeding appendages attaches near the front,

3.65. A recent reconstruction of the Burgess Shale fauna (Conway Morris and Whittington, 1985), showing the new interpretation of *Anomalocaris* (24), and the great size of this creature compared to the others. Note the weird wonders *Opabinia* (8), *Dinomischus* (9), and *Wiwaxia* (23); and the arthropods *Aysheaia* (5), *Leanchoilia* (6), *Yohoia* (11), *Canadaspis* (12), *Marrella* (15), and *Burgessia* (19).

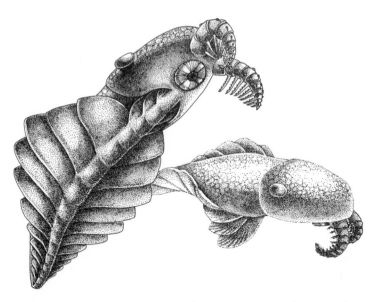

3.66. The two known species of *Anomalocaris:* top, *Anomalocaris nathorsti* as seen from below, showing the circular mouth, misidentified by Walcott as a jellyfish, and the pair of feeding appendages; bottom, *Anomalocaris canadensis* as seen from the side, in swimming position. Drawn by Marianne Collins.

with the circlet of the mouth behind and in the mid-line (figure 3.67). The plates of the circlet could substantially constrict the area of the mouth but not fully come together (in any orientation that Whittington or Briggs could reconstruct), so the mouth probably remained permanently open, at least partially. Whittington and Briggs conjecture that the mouth may have worked like a nutcracker, with *Anomalocaris* using its appendages to bring prey to the opening (figure 3.68), and then crushing its food by constriction. The inner borders of the plates in the *Peytoia* circlet all bear teeth. In one specimen, Whittington and Briggs found three additional rows of teeth, stacked one above the other parallel to the circlet of mouth plates. The teeth in these rows may have been attached to the circlet, but they probably extended from the walls of the gullet—thus providing *Anomalocaris* with a formidable array of weapons both in the mouth itself and in the front end of the gut (figure 3.69).

Behind the mouth at the ventral surface, the head carries three pairs of strongly overlapping lobes (see figure 3.67). The trunk behind the head is divided into eleven lobes, each triangular in basic shape, with the apex

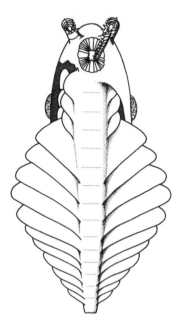

3.67. *Anomalocaris* as seen from below, showing how the feeding appendages could bring food to the mouth (Whittington and Briggs, 1985). Just behind the mouth at the left, part of the ventral surface of the animal has been omitted, to show the gills lying above the three posterior segments of the head.

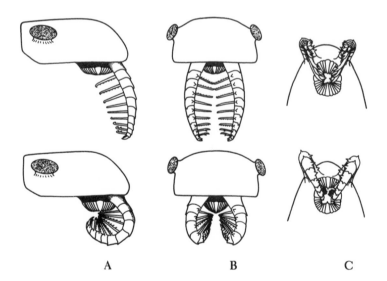

A   B   C

3.68. The probable mode of feeding of *Anomalocaris*. (A) The head of *Anomalocaris nathorsti* seen from the side, with the feeding appendage extended (top) and coiled up to bring food to the mouth (bottom). (B) The same operation viewed from the front. (C) As seen from below, the feeding appendage coiled to bring food to the mouth, in *Anomalocaris nathorsti* (top) and in *Anomalocaris canadensis* (bottom).

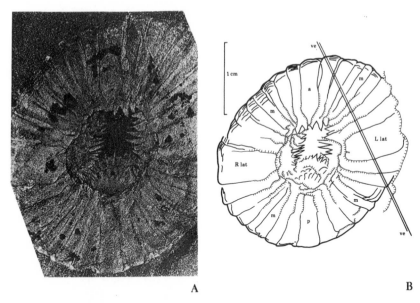

A   B

3.69. The mouth of *Anomalocaris,* mistaken by Walcott for the jellyfish
Peytoia. Several rows of teeth can be seen extending down from the central
space; these tooth rows may be projecting from the gullet of the animal.
(A) A photograph of the specimen. (B) A camera lucida drawing of the same
specimen.

pointed back in the mid-line. The lobes are widest at the middle of the
trunk, evenly tapering both in front and behind. These lobes, like the three
at the rear of the head, strongly overlap. The termination of the trunk is
short and blunt, without any projecting spine or lobe. A multilayered structure of stacked lamellae, presumably a gill, attaches to the top surface of
each lobe.

Since *Anomalocaris* has no body appendages, it presumably did not walk
or crawl along the substrate. Whittington and Briggs reconstruct
Anomalocaris as a capable swimmer, though no speed demon, propelled by
wavelike motions of the body lobes in coordinated sequences (figure 3.70).
The overlapping lateral lobes would therefore work much like the single
lateral fin flap of some fishes. An *Anomalocaris* in motion may have resem-
bled a modern manta ray, undulating through the water by generating
waves within the broad and continuous fin.

Again, as with *Wiwaxia* and *Opabinia,* one can make reasonable conjec-
tures about the biological operation of *Anomalocaris*—a creature can,
after all, only eat and move in so many ways. But what could such an odd

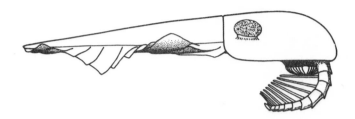

3.70. Reconstruction of *Anomalocaris* as seen from the side, in the act of swimming (Whittington and Briggs, 1985).

animal be in genealogical terms? The feeding appendages had been read as arthropod parts for a century—and their segmented character does recall the great phylum of joint-footed creatures. But repetition and segmentation, shown by the sequence of lobes as well as the feeding appendages, are not restricted to arthropods—think of annelids, vertebrates, and even the molluscan "living fossil," *Neopilina.* Nothing else about *Anomalocaris* suggests a linkage with arthropods. The body bears no jointed appendages, and the mouth, with its perpetual gape and circlet of plates, is unique, utterly unlike anything in the phylum Arthropoda. Even the pair of feeding appendages, though segmented, strays far from any arthropod prototype as soon as we attempt any comparison in detail. Whittington and Briggs concluded that *Anomalocaris* "was a metameric animal, and had one pair of jointed appendages and a unique circlet of jaw plates. We do not consider it an arthropod, but the representative of a hitherto unknown phylum" (1985, p. 571).

CODA

The Burgess work will continue, for many genera remain ripe for restudy (the bulk of the arthropods have been monographed, but only about half of the known weird wonders). However, Harry, Derek, and Simon are moving on, for various reasons. The Lord gives us so little time for a career—forty years if we start early as graduate students and remain in good health, fifty if fortune smiles. The Devil takes so much away—primarily in administrative burdens that fall upon all but the most resistant and singularly purposeful of SOBs. (The earthly rewards of scholarship are higher offices that extinguish the possibility of future scholarship.) You can't spend an entire career on one project, no matter how important or exciting. Harry, in his seventies, has returned to his first love, and is spearheading a revision of the trilobite volume for the *Treatise on Invertebrate Paleontology.* Simon's burgeoning career includes a Burgess Shale project or two, but his main interests have moved backward in time to the Cambrian explosion itself. Derek's expanding concerns center on weird wonders and soft-bodied faunas of post-Burgess times.

Others will finish this generation's run at the Burgess Shale. And then the next generation will arrive with new ideas and new techniques. But science is cumulative, despite all its backings and forthings, ups and downs. The work of Briggs, Conway Morris, and Whittington will be honored for its elegance and for the power of its transforming ideas as long as we maintain that most precious of human continuities—an unbroken skein of intellectual genealogy.

No organism or interpretation can have the last word in such a drama, but we must respect the closure of a man's work. The epilogue to this play belongs to Harry Whittington, who in his typically succinct and direct words, wrote to me about his Burgess monographs: "Perhaps these necessarily dry papers conveyed a little of the excitement of discovery—it certainly was an intriguing investigation which had its moments of great joy when a new and unexpected structure was revealed by preparation" (March 1, 1988). "It has been the most exciting and intriguing project that I have been associated with" (April 22, 1987).

Summary Statement on the Bestiary of the Burgess Shale

DISPARITY FOLLOWED BY DECIMATION: A GENERAL STATEMENT

If the soft-bodied components had never been found, the Burgess Shale would be an entirely unremarkable Middle Cambrian fauna of about thirty-three genera. It contains a rich assemblage of sponges (Rigby, 1986) and algae, seven species of brachiopods, nineteen species of ordinary trilobites with hard parts, four of echinoderms, and a mollusk and coelenterate or two (Whittington, 1985b, pp. 133–39, presents a complete list). Among the soft-bodied organisms, bringing the total biota to about 120 genera, some are legitimate members of major groups. Whittington lists five certain and two probable species of priapulid worms, six species of polychaetes, and three soft-bodied trilobites (*Tegopelte* and two species of *Naraoia*).

My five-act drama, just concluded, emphasizes a different theme, taught to me by the soft-bodied components alone. The Burgess Shale includes a range of disparity in anatomical design never again equaled, and not matched today by all the creatures in all the world's oceans. The history of multicellular life has been dominated by decimation of a large initial stock, quickly generated in the Cambrian explosion. The story of the last 500 million years has featured restriction followed by proliferation within a few stereotyped designs, not general expansion of range and increase in complexity as our favored iconography, the cone of increasing diversity, implies. Moreover, the new iconography of rapid establishment and later decimation dominates all scales, and seems to have the generality of a fractal pattern. The Burgess revisions of Whittington and colleagues have specified three ascending levels.

1. *Major groups of a phylum.* No group of invertebrate fossils has received more study, or stands higher in general popularity, than trilobites. The mineralized skeletons of conventional fossils show extraordinary diversity, but all conform to a basic design. One would hardly have anticipated, after all this study, that the total anatomical range of the group could have been far broader in its early days. Yet soft-bodied *Naraoia* is undoubtedly a trilobite in its distinctive series of head appendages (one pair of antennae and three post-oral biramous pairs), and its conventional body appendages of the "right" form and number of segments. Yet the exoskeleton of *Naraoia*, with its two valves, stands far outside the anatomical range of the group as seen in conventional fossils.

2. *Phyla.* We can completely grasp the extent of a surprise only when we also know the full range of conventional possibilities—for we need a baseline of calibration. I find the story of Burgess arthropods particularly satisfying because the baseline has "no vacancy," and all additional disparity truly supplements a full range of membership in major groups. The orphaned arthropods of the Burgess are spectacular, but the representatives of conventional groups are just as important for documenting the first phrase of the primary theme—"all we could expect and then a great deal more." The recent discovery of *Sanctacaris* brings the conventional roster to completion. All four great groups of arthropods have representatives in the Burgess Shale:

Trilobita—nineteen ordinary species plus three soft-bodied
Crustacea—*Canadaspis* and perhaps *Perspicaris*
Uniramia—*Aysheaia*, if correctly identified as an onychophoran
Chelicerata—*Sanctacaris*

But the Burgess Shale contains an even greater range of anatomical experiments, equally distinct in design and functionally able, but not leading to subsequent diversity. A few of these orphans may show relationships among themselves—*Actaeus* and *Leanchoilia*, perhaps, on the basis of their distinctive frontal appendages—but most are unique, with defining features shared by no other species.

The monographic work of Whittington and colleagues has identified thirteen unique designs (table 3.3), all discussed in the preceding chronology. But how many more have yet to be described? Whittington lists twenty-two species (and inadvertently omits *Marrella*) in his category "not placed in any phylum or class of Arthropoda" (1985b, p. 138). Therefore, by best estimate, the Burgess Shale contains at least twenty unique designs of arthropods, in addition to the documented representatives of all four great groups within the phylum.*

3. *Multicellular animal life as a whole.* The weird wonders of the Burgess Shale excite our greatest fascination, though the arthropod story is every bit as satisfying intellectually, especially for its completion of the baseline and consequently firm estimate for the relative frequency of odd-

---

*If I wished to play devil's advocate against my own framework, I would argue that the criterion by which we make the claim of twenty losers and only four winners is falsely retrospective. By patterns of tagmosis, modern arthropods are surely strikingly less disparate than Burgess forebears. But why use patterns of tagmosis as a basis for higher-level classification of arthropods? A nearly microscopic ostracode, a terrestrial isopod, a planktonic copepod, a Maine lobster, and a Japanese king crab span more variety in size and ecological specialization than all the Burgess arthropods put together—though all these modern creatures are called Crustacea, and display the stereotyped tagmosis of this class. A paleontologist living during the Burgess might consider the arthropods as less varied because he had no reason to regard patterns of tagmosis as a particularly important character (for the utility of tagmosis in distinguishing major genealogical lines only became apparent later, after most alternatives were decimated and stereotypy set in among the few surviving and highly disparate lines).

I regard this argument as a poor case. If you wish to reject tagmosis as too retrospective then what other criterion will suggest less disparity in the Burgess? We use basic anatomical designs, not ecological diversification, as our criterion of higher-level classification (bats and whales are both mammals). Nearly every Burgess genus represents a design unto itself by any anatomical criterion. Tagmosis does stabilize in post-Burgess times, as do arrangement and forms of appendages—while no major feature of arthropod design can distinguish broad and stable groups in the Burgess.

TABLE 3.3. *The Burgess Drama: Dramatis Personae in Order of Appearance*

	Year of Redescription	Name	Status for Walcott	Status As Revised	Reviser
Act 1	1971	*Marrella*	close to Trilobita	unique arthropod	Whittington
	1974	*Yohoia*	branchiopod crustacean	unique arthropod	Whittington
	1975	*Olenoides*	trilobite (called *Nathorstia*)	trilobite	Whittington
Act 2	1975	*Opabinia*	branchiopod crustacean	new phylum	Whittington
Act 3	1975	*Burgessia*	branchiopod crustacean	unique arthropod	Hughes
	1976	*Nectocaris*	(unknown)	new phylum	Conway Morris
	1976	*Odontogriphus*	(unknown)	new phylum	Conway Morris
	1977	*Dinomischus*	(unknown)	new phylum	Conway Morris
	1977	*Amiskwia*	chaetognath worm	new phylum	Conway Morris
	1977	*Hallucigenia*	polychaete worm	new phylum	Conway Morris
	1976	*Branchiocaris*	malacostracan crustacean	unique arthropod	Briggs
	1976	*Perspicaris*	malacostracan crustacean	(?) malacostracan	Briggs
	1978	*Canadaspis*	malacostracan crustacean (called *Hymenocaris*)	malacostracan	Briggs
Act 4	1977	*Naraoia*	branchiopod crustacean	soft-bodied trilobite	Whittington
	1985	*Tegopelte*	(unknown)	soft-bodied trilobite	Whittington
	1978	*Aysheaia*	polychaete worm	(?) onychophoran or new phylum	Whittington

TABLE 3.3. *The Burgess Drama: Dramatis Personae in Order of Appearance (Continued)*

	Year of Redescription	Name	Status for Walcott	Status As Revised	Reviser
Act 5	1981	*Odaraia*	malacostracan crustacean	unique arthropod	Briggs
	1981	*Sidneyia*	merostome	unique arthropod	Bruton
	1981	*Molaria*	merostome	unique arthropod	Whittington
	1981	*Habelia*	merostome	unique arthropod	Whittington
	1981	*Sarotrocercus*	(unknown)	unique arthropod	Whittington
	1981	*Actaeus*	(unknown)	unique arthropod	Whittington
	1981	*Alalcomenaeus*	(unknown)	unique arthropod	Whittington
	1983	*Emeraldella*	merostome	unique arthropod	Bruton and Whittington
	1983	*Leanchoilia*	branchiopod crustacean	unique arthropod	Bruton and Whittington
	1988	*Sanctacaris*	(unknown)	chelicerate arthropod	Briggs and Collins
	1985	*Wiwaxia*	polychaete worm	new phylum	Conway Morris
	1985	*Anomalocaris*	branchiopod crustacean	new phylum	Whittington and Briggs
		(*Laggania*)	sea cucumber	body of *Anomalocaris*	
		(*Peytoia*)	jellyfish	mouth of *Anomalocaris*	
		(Appendage F)	feeding limb of *Sidneyia*	feeding organ of *A. nathorsti*	

balls. Still, whereas *Marrella* and *Leanchoilia* may be beautiful and surprising, *Opabinia*, *Wiwaxia*, and *Anomalocaris* are awesome—deeply disturbing and thrilling at the same time.

The Burgess revision has identified eight anatomical designs that do not fit into any known animal phylum: in order of publication, *Opabinia*, *Nectocaris*, *Odontogriphus*, *Dinomischus*, *Amiskwia*, *Hallucigenia*, *Wiwaxia*, and *Anomalocaris*. But this list is nowhere near complete—surely less exhaustive than the account of documented oddballs among arthropods. The best estimates indicate that only about half the weird wonders of the Burgess Shale have been described. Two recent sources have provided lists of all potential creatures in this category of ultimate strangeness. Whittington counts seventeen species of "miscellaneous animals" (1985b, p. 139), and I would add *Eldonia* to his total. Briggs and Conway Morris count nineteen "Problematica from the Middle Cambrian Burgess Shale of British Columbia" (1986). Finding no basis for genealogical or anatomical arrangement among the weird wonders, they simply list their nineteen creatures in alphabetical order.

What may the future bring us in further surprises from the Burgess Shale? Consider *Banffia*, namesake of the more famous national park adjoining Yoho and the Burgess Shale. Walcott's "worm"—with an annulated front portion separated from a saclike posterior—is almost surely a weird wonder. Or *Portalia*, an elongate animal with bifurcating tentacles arrayed along the body axis. Or *Pollingeria*, a scalelike object with a meandering tubelike structure on top. Walcott interpreted *Pollingeria* as a covering plate from a larger organism, akin to the sclerites of *Wiwaxia*, and explained the meandering tube as a commensal worm, but Briggs and Conway Morris think that the object could be an entire organism. The general form of the Burgess story may now be well in hand, but Walcott's quarry has not yet yielded all its particular treasures.

ASSESSMENT OF GENEALOGICAL RELATIONSHIPS FOR BURGESS ORGANISMS

This book, long enough already, cannot become an abstract treatise on the rules of evolutionary inference. But I do need to provide a few explicit comments on how paleontologists move from descriptions of anatomy to proposals about genealogical relationships—so that my numerous statements on this subject receive some underpinning and do not stand as undefended pronouncements *ex cathedra*.

Louis Agassiz, the great zoologist who founded the institution that now houses both me and the Raymond collection of Burgess Shale fossils, picked a superficially peculiar name that we retain with pride—the Museum of Comparative Zoology. (Anticipating the hagiographical urges of his contemporaries, he even explicitly requested that his chosen title be retained in perpetuity, and that the museum not be renamed for him upon his demise.) Experiment and manipulation may form the stereotype of science, Agassiz argued, but disciplines that treat the inordinately complex, unrepeatable products of history must proceed differently. Natural history must operate by analyzing similarities and differences within its forest of unique and distinctive products—in other words, by comparison.

Evolutionary and genealogical inferences rest upon the study and meaning of similarities and differences, and the basic task is neither simple nor obvious. If we could just compile a long list of features, count the likenesses and unlikenesses, gin up a number to express an overall level of resemblances, and then equate evolutionary relationship with measured similarity, we could almost switch to automatic pilot and entrust our basic job to a computer.

The world, as usual, is not so simple—and thank goodness, for the horizon would probably be a disappointing place anyway. Similarities come in many forms: some are guides to genealogical inferences; others are pitfalls and dangers. As a basic distinction, we must rigidly separate similarities due to simple inheritance of features present in common ancestors, from similarities arising by separate evolution for the same function. The first kind of similarity, called homology, is the proper guide to descent. I have the same number of neck vertebrae as a giraffe, a mole, and a bat, not (obviously) because we all use our heads in the same way, but because seven is the ancestral number in mammals, and has been retained by descent in nearly all modern groups (sloths and their relatives excepted). The second kind of similarity, called analogy, is the most treacherous obstacle to the search for genealogy. The wings of birds, bats, and pterosaurs share some basic aerodynamic features, but each evolved independently; for no common ancestor of any pair had wings. Distinguishing homology from analogy is the basic activity of genealogical inference. We use a simple rule: rigidly exclude analogies and base genealogies on homology alone. Bats are mammals, not birds.

Using this cardinal rule, we can go a certain distance with the Burgess Shale. The tail flukes of *Odaraia* bear an uncanny resemblance to functionally similar structures of some fishes and marine mammals. But *Odaraia* is clearly an arthropod, not a vertebrate. *Anomalocaris* may have used its overlapping lateral flaps to swim by undulation, much as certain fishes with continuous lateral fins or flattened body edges do—but this functional similarity, evolved from different anatomical foundations, indicates nothing about genealogical relationship. *Anomalocaris* remains a weird wonder, no closer to a vertebrate than to any other known creature.

But the basic distinction between homology and analogy will not carry us far enough. We must make a second division, among homologous structures themselves. Rats and people share both hair and a vertebral column. Both are homologies, structures inherited from common ancestors. If we are searching for a criterion that will properly unite rats and people into the genealogical group of mammals, we can use hair, but the shared vertebral column will not help us at all. Why the difference? Hair works because it is a *shared-and-derived* character, confined to mammals among the vertebrates. A vertebral column is no help because it is a *shared-but-primitive* character, present in the common ancestor of all terrestrial vertebrates—not just mammals—and most fish.

This distinction between properly restricted (shared and derived) and overly broad homologies (shared but primitive) lies at the core of our greatest contemporary difficulties with Burgess organisms.* For example, many Burgess arthropods have a bivalved carapace; many others share the basic "merostomoid" form, a broad head shield followed by numerous short and wide body segments capped by a tail spike. These two features are, presumably, genuine arthropod homologies—each bivalved lineage doesn't start from scratch and develop the same complex structure, slowly and separately. But neither the presence of a bivalved carapace nor "merostomoid" body form can identify a genealogically coherent group of Burgess arthropods because both are shared-but-primitive characters.

Figure 3.71 should clarify the reason for rejecting shared-but-primitive traits as a guide to genealogy. This evolutionary tree represents a lineage that has diversified into three great groups—I, II, and III—by the time marked by the dashed line. A star indicates the presence of a homologous trait—call it five digits on the front limb—inherited from the distant common ancestor (A). In many branches, this trait has been lost or modified beyond recognition. Every loss is marked by a double-headed arrow. Note that at the selected time, four species (1–4) still retain the shared-but-primitive trait. If we united these four as a genealogical group, we would be making the worst possible error—missing the three true groups entirely, while taking members from each to construct a false assemblage: species 1 might be the ancestor of horses; species 2 and 3, early rodents; and species 4, an ancestor of primates, including humans. The fallacy of

---

*Many of Walcott's cruder errors, on the other hand—confusing the sclerites of *Wiwaxia* with setae of polychaetes, and the lateral flaps of *Opabinia* with arthropod segments—represented a more basic failure to distinguish analogy from homology.

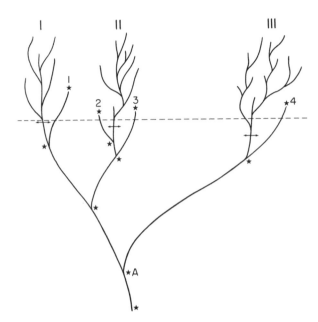

3.71.   A hypothetical evolutionary tree illustrating why shared-but-primitive traits must be rejected as guides in identifying genealogical groups. Lineages and branching points marked with a star possess the shared-but-primitive trait. Double-headed arrows mark the loss of this trait.

basing groups on shared-but-primitive traits should be apparent.*

But the Burgess problem is probably even worse. In my five-act chronology, I often spoke of a grabbag of available arthropod characters. Suppose that such shared-but-primitive features as the bivalved carapace, unlike the starred trait of figure 3.71, do not indicate continuous lineages. Suppose that in this early age of unparalleled experimentation and genetic lability, such traits could arise, again and again, in any new arthropod lineage—not by slow and separate evolution for common function (for the traits would

*Thus, we can take some steps to resolving the genealogy of Burgess organisms. We can eliminate some resemblances based on analogy—setae of polychaetes and sclerites of *Wiwaxia*, for example. We can also eliminate some shared-but-primitive characters that do not define genealogical groups—bivalved carapaces and "merostomoid" body form. But the identification of shared-and-derived characters has been largely unsuccessful so far. Homology of shared-and-derived frontal appendages may unite *Leanchoilia* with *Actaeus* (and perhaps also with *Alalcomenaeus*). The lateral flaps with gills above may be shared-and-derived characters in *Opabinia* and *Anomalocaris*, thus constituting the only genealogical linkage between two of the weird wonders.

then represent classic analogies), but as latent potentials in the genetic system of all early arthropods, separately recruitable for overt expression in each lineage. Then traits like merostomoid body form and bivalved carapaces would pop up again and again all over the arthropod evolutionary tree.

I suspect that such a strange phenomenon did prevail in Burgess times, and that we have had so little success in reconstructing Burgess genealogies because each species arose by a process not too different from constructing a meal from a gigantic old-style Chinese menu (before the Szechuan, yuppie, and other gastronomical revolutions)—one from column A, two from B, with many columns and long lists in each column. Our ability to recognize coherent groups among later arthropods arises for two reasons: First, lineages lost this original genetic potential for recruitment of each major part from many latent possibilities; and second, the removal of most lineages by extinction left only a few survivors, with big gaps between (figure 3.72). The radiation of these few surviving lineages (into a great diversity of species with restricted disparity of total form) produced the distinct

3.72. A hypothetical evolutionary tree reflecting a view of life's history suggested by the reinterpretation of the Burgess fauna. The removal of most groups by extinction leaves large morphological gaps among the survivors. The dashed line represents the time of the Burgess Shale, with disparity at a maximum.

groups that we know today as phyla and classes.

I think that Derek Briggs had a model like this in mind when he wrote of the difficulty in classifying Burgess arthropods: "Each species has unique characteristics, while those shared tend to be generalized and common to many arthropods. Relationships between these contemporaneous species are, therefore, far from obvious, and possible ancestral forms are unknown." (1981b, p. 38).*

I also think that the model of the grabbag might be extended to all Burgess animals taken together, not only to the arthropods separately. What are we to make of the feeding appendages on *Anomalocaris?* They do seem to be fashioned on an arthropod plan, but the rest of the body suggests no affinity with this great phylum. Perhaps they are only analogous to arthropod limbs, separately evolved and truly devoid of any genetic continuity with the jointed structures of arthropods. But perhaps the Burgess grabbag extended across phyla. Perhaps jointed structures with a common genetic underpinning were not yet restricted to the Arthropoda. Their limited presence elsewhere would not imply close genealogical relationship with arthropods, but only a broad range of latent and recruitable structures that did not yet respect the later, unbridgeable boundaries of modern phyla. The jaws of *Wiwaxia* (recalling the molluscan radula) and the feeding organ of *Odontogriphus* (recalling the lophophore of several phyla) come to mind as other possible features from the mega-grabbag.

The model of the grabbag is a taxonomist's nightmare and an evolutionist's delight. Imagine an organism built of a hundred basic features, with twenty possible forms per feature. The grabbag contains a hundred compartments, with twenty different tokens in each. To make a new Burgess creature, the Great Token-Stringer takes one token at random from each compartment and strings them all together. *Voilà,* the creature works—and you have nearly as many successful experiments as a musical scale can build catchy tunes.† The world has not operated this way since Burgess

---

*Technical footnote: Several efforts have been made to construct a cladogram for the Burgess arthropods (Briggs, 1983, and in press). These have, so far, been conspicuously unsuccessful, as the different possibilities do not satisfactorily converge. If the grabbag model is correct, and each major feature of each new lineage arises separately from a suite of latent possibilities common to all, then genealogical connectivity of phenotypes is broken, and the problem may be intractable by ordinary cladistic methods. Of course, some continuity in some genuinely nested sets of characters may well exist, but the appropriate features will be difficult to identify.

†I exaggerate to make a point. Rules of construction and order pervade nature. Not all conceivable combinations can work, nor can all amalgams be constructed within the developmental constraints of metazoan embryology. I use this metaphor only to express the vastly expanded range of Burgess possibilities.

times. Today, the Great Token-Stringer uses a variety of separate bags—labeled "vertebrate body plan," "angiosperm body plan," "molluscan body plan," and so forth. The tokens in each compartment are far less numerous, and few if any from bag 1 can also be found in bag 2. The Great Token-Stringer now makes a much more orderly set of new creatures, but the playfulness and surprise of his early work have disappeared. He is no longer the *enfant terrible* of a brave new multicellular world, fashioning *Anomalocaris* with a hint of arthropod, *Wiwaxia* with a whiff of mollusk, *Nectocaris* with an amalgam of arthropod and vertebrate.

The story is old, and canonical. The youthful firebrand has become the apostle of good sense and stable design. Yet the former spark is not entirely extinct. Something truly new slips by now and then within the boundaries of strict inheritance. Perhaps his natural vanity finally got the better of him. Perhaps he couldn't bear the thought of running such an exquisite play for so long, and having no chronicler to admire the work. So he let the token for more brain tumble from compartment 1 of the primate bag—and assembled a species that could paint the caves of Lascaux, frame the glass of Chartres, and finally decipher the story of the Burgess Shale.

THE BURGESS SHALE AS A CAMBRIAN GENERALITY

The chief fascination of the Burgess Shale lies in a paradox of human comprehension. The most stunning and newsworthy parts of the story involve the greatest oddities and strangest creatures. *Anomalocaris*, two feet long, and crunching a trilobite in its circular "jellyfish" jaw, rightly wins the headlines. But the human mind needs anchors in familiarity. The Burgess teaches us a general lesson, and reverses our usual view of life, because so much about this fauna has the clear ring of conventionality. Its creatures eat and move in ordinary ways; the entire community strikes a working ecologist as comprehensible in modern terms; key elements of the fauna also appear in other locations, and we learn that the Burgess represents the normal world of Cambrian times, not a bizarre marine grotto in British Columbia.

I emphasized throughout my five-act chronology that the discovery of conventional creatures, true crustaceans and chelicerates, was every bit as important as the reconstruction of weird wonders in forging a complete interpretation for the Burgess Shale. If we now take a larger look, and consider the entire fauna as a totality, as a functioning ecological community, the same theme holds with even more force. The anatomical oddness of the Burgess gains its meaning against a backdrop of global spread and conventional ecology for the fauna as a whole.

PREDATORS AND PREY: THE FUNCTIONAL WORLD OF BURGESS ARTHROPODS

In 1985, Briggs and Whittington published a fascinating article summarizing their conclusions on the modes of life and ecology of Burgess arthropods; (the focus of almost all their previous monographic work had been anatomical and genealogical). Taking all the arthropods together, they inferred a range of behaviors and feeding styles comparable with modern faunas. They divided the Burgess genera into six major ecological categories.

1. *Predatory and scavenging benthos.* (Benthic creatures live on the sea floor and do little or no swimming.) This large group includes the trilobites and several of the "merostomoid" genera—*Sidneyia, Emeraldella, Molaria,* and *Habelia* (figure 3.73D and F–K). All have biramous body appendages bearing strong walking branches with a spiny inner border on the first segment, facing the central food groove. The alimentary canal (where identified) curves down and backward at the mouth—indicating that food was passed from the rear forward, as in most benthic arthropods. The strong spines imply that relatively large food items were caught or scavenged, and passed forward to the mouth.

2. *Deposit-feeding benthos.* (Deposit feeders extract small particles from sediment, often by processing large quantities of mud; they do not select or actively pursue large food items.) Several genera fall into this category, primarily on the evidence of weak or absent spines on the inner borders of the food groove—*Canadaspis, Burgessia, Waptia,* and *Marrella,* for example (figure 3.74E and H–J). Most of these genera could probably either walk across the bottom sediment or swim weakly in the water column just above.

3. *Scavenging, and perhaps predatory, nektobenthos.* (Nektobenthonic creatures both swim and walk on the sea floor.) The genera in this category—*Branchiocaris* and *Yohoia* (figure 3.74D and F)—were not primarily benthic because they did not possess biramous appendages with strong walking branches. *Yohoia* has three biramous appendages on the head, but probably uniramous limbs with gill branches, used for respiration and swimming, alone on the body; *Branchiocaris* has biramous body appendages, but with short, weak walking branches. The absence of strong inner branches on the body appendages also suggests that these genera did not eat by passing food forward from the rear. But both genera possess large

head appendages with claws at the tip, and probably brought discrete food items from the front end of the body directly to the mouth.

4. *Deposit-feeding and scavenging nektobenthos.* Like the genera of the preceding category, the members of this group have body appendages with weak or absent inner branches, implying little walking and food processing from the rear; stronger outer branches for swimming; and head appendages that could have gathered food directly. But these genera—*Leanchoilia*,

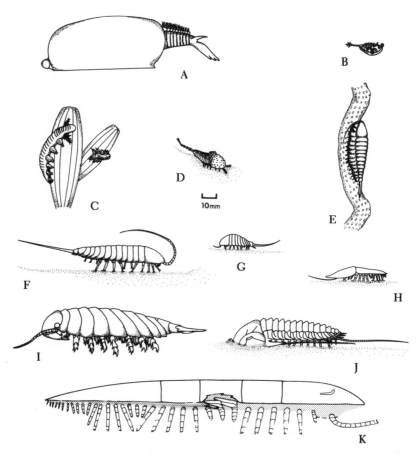

3.73. Burgess arthropods, all drawn to the same scale to show their relative sizes (Briggs and Whittington, 1985). (A) *Odaraia.* (B) *Sarotrocercus.* (C) *Aysheaia.* (D) *Habelia.* (E) *Alalcomenaeus.* (F) *Emeraldella.* (G) *Molaria.* (H) *Naraoia.* (I) *Sidneyia.* (J) The trilobite *Olenoides.* (K) The large soft-bodied trilobite *Tegopelte.*

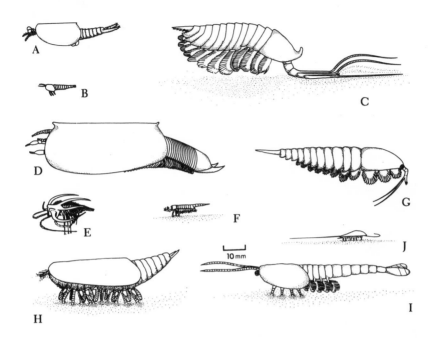

3.74. From Briggs and Whittington, 1985. Additional Burgess arthropods, all drawn to the same scale. (A) *Perspicaris*. (B) *Plenocaris*. (C) *Leanchoilia*. (D) *Branchiocaris*. (E) *Marrella*. (F) *Yohoia*. (G) *Actaeus*. (H) *Canadaspis*. (I) *Waptia*. (J) *Burgessia*.

Actaeus, Perspicaris, and *Plenocaris* (figure 3.74A–C and G)—do not have strong claws on the tips of their frontal appendages, and probably did not capture large food items; hence they are regarded as probable deposit feeders.

5. *Nektonic suspension feeders.* This small category—consisting of *Odaraia* and *Sarotrocercus* (figure 3.73A–B)—includes the true swimmers among Burgess arthropods. These genera either had no walking branches *(Sarotrocercus)* or possessed short inner branches that could not extend beyond the carapace *(Odaraia).* They had the biggest eyes among Burgess arthropods, and both probably sought small prey for filter feeding.

6. *Others.* Every classification has a residual category for unusual members. *Aysheaia* (figure 3.73C) may have been a parasite, living among and feeding on sponges. *Alalcomenaeus* (figure 3.73E) bears strong spines all along the inner edges of its walking legs, not only on the first segment, adjoining the food groove. Briggs and Whittington conjecture that *Alal-*

comenaeus may have used these spines either to grasp on to algae, or to tear carcasses in scavenging.

Briggs and Whittington include two excellent summary figures in their paper (figures 3.73 and 3.74). Each genus is shown in its probable habitat, and all are drawn to the same scale—so that the substantial differences in size among genera may be appreciated.

Each of the six categories crosses genealogical lines. The ensemble fills a set of ordinary roles for modern marine arthropods. The great anatomical disparity among Burgess arthropods is therefore not a simple adaptive response to a wider range of environments available at this early time. Somehow, the same basic scope of opportunity originally elicited a far greater range of anatomical experimentation. Same ecological world; very different kind of evolutionary response: this situation defines the enigma of the Burgess.

THE ECOLOGY OF THE BURGESS FAUNA

In 1986, a year after his monograph on *Wiwaxia,* Simon Conway Morris published a "blockbuster" of another type—a comprehensive ecological analysis of the entire Burgess community. He began with some interesting facts and figures. About 73,300 specimens on 33,520 slabs have been collected from the Burgess Shale. Ninety percent of this material resides in Washington, in Walcott's collection; 87.9 percent of these specimens are animals, and nearly all the rest are algae. Fourteen percent of the animals have shelly skeletons; the remainder are soft-bodied.

The fauna contains 119 genera in 140 species; 37 percent of these genera are arthropods. Conway Morris identified two main elements in the fauna: (1) An overwhelmingly predominant assemblage of benthic and near-bottom species that were transported into a stagnant basin by the mudslide. Conway Morris inferred, from abundant algae needing light for photosynthesis, that this assemblage originally lived in shallow water, probably less than three hundred feet in depth. He called this element the *Marrella-Ottoia* assemblage, to honor both the most common substrate walker (the arthropod *Marrella*) and the most common burrower (the priapulid worm *Ottoia*). (2) A much rarer group of permanently swimming creatures that lived in the water column above the stagnant basin, and settled amidst the animals transported by the mudslide. Conway Morris called this element the *Amiskwia-Odontogriphus* assemblage, to honor two of his pelagic weird wonders.

He found that the Burgess genera, despite their odd and disparate anatomies, fall into conventional categories when classified by feeding style and habitat. He recognized four major groups: (1) Deposit-feeding collectors (mostly arthropods)—60 percent of the total number of individuals; 25–30 percent of the genera. (This category includes *Marrella* and *Canadaspis*, the two most common Burgess animals, hence the high representation for individuals). (2) Deposit-feeding swallowers (mostly ordinary mollusks with hard parts)—1 percent of individuals; 5 percent of genera. (3) Suspension feeders (mostly sponges, taking food directly from the water column)—30 percent of individuals; 45 percent of genera. (4) Carnivores and scavengers (mostly arthropods)—10 percent of individuals; 20 percent of genera.

Traditional wisdom, with its progressionist bias and its iconography of the cone of increasing diversity, has viewed Cambrian communities as more generalized and less complex than their successors. Cambrian faunas have been characterized as ecologically unspecialized, with species occupying broad niches. Trophic structure has been judged as simple, with detritus and suspension feeders dominating, and predators either rare or entirely absent. Communities have been reconstructed with broad environmental tolerances, large geographic distributions, and diffuse boundaries.

Conway Morris did not entirely overturn these received ideas of a relatively simple world. He did, for example, find comparatively little complexity in the attacking and maneuvering capacities of Burgess predators: "It seems plausible that the degree of sophistication in styles of predation (search and attack) and deterrence in comparison with younger Paleozoic faunas was substantially less" (1986, p. 455).

Still, his primary message made the ecology of the Burgess Shale more conventional, and more like the worlds of later geological periods. Over and over again, when the full range of this community could be judged by its soft-bodied elements, Conway Morris found more richness and more complexity than earlier views had allowed. Detritus and suspension feeders did dominate, but their niches did not overlap broadly, with all species simply sopping up everything edible in sight. Rather, most organisms were specialized for feeding on particular types and sizes of food in a definitely limited environment. Suspension feeders did not absorb all particles at all levels in the water column; the various species were, as in later faunas, "tiered" in assemblages of complex interaction. (In tiering, various forms specialize, confining themselves to low, medium, or high level of the water column, as communities diversify.) Most surprising of all, predators played a major role in the Burgess community. This top level of the ecological pyramid was fully occupied and functioning. No longer could the disparity of early form be attributed to reduced pressures of an easy world, devoid of

Darwinian competition in the struggle for existence, and therefore open to any contraption or jury-rigged experiment. The Burgess fauna, Conway Morris argued, "shows unequivocally that the fundamental trophic structure of marine metazoan life was established early in its evolution" (1986, p. 458).

Conway Morris had reached the same conclusion for the entire Burgess ecology that Briggs and Whittington had established for arthropod life styles. The "ecological theater" of the Burgess Shale had been rather ordinary: "It may transpire," Conway Morris wrote, "that the community structure of the Phyllopod Bed was not fundamentally different from that of many younger Paleozoic soft-bodied faunas" (1986, p. 451). Why then was the "evolutionary play" of these early times so different?

THE BURGESS AS AN EARLY WORLD-WIDE FAUNA

Nothing breeds scientific activity quite so effectively as success. The excitement generated by recent work on the Burgess Shale has inspired an outburst of interest in soft-bodied faunas and the history of early multicellular life. The Burgess Shale is a small quarry in British Columbia, deposited in Middle Cambrian times, after the celebrated explosion of the Lower Cambrian. As long as its fauna remained geographically confined, and temporally limited to a mere moment after the main event, the Burgess Shale could not tell a story for all of life. The most exciting development of the past decade, continuing and accelerating as I write this book, lies in the discovery of Burgess genera all over the world, and in earlier rocks.

The first and most obvious extension occurred close to home. If a mudslide down an unstable slope formed the Burgess, many other slides must have occurred in adjacent regions at about the same time; some must have been preserved. As previously discussed, Des Collins of the Royal Ontario Museum has pioneered the effort to find these Burgess equivalents, and he has been brilliantly successful; during the 1981 and 1982 field seasons, Collins found more than a dozen Burgess equivalents in areas within twenty miles or less of the original site. Briggs and Conway Morris joined the field party in 1981, and Briggs returned in 1982. (See Collins, 1985; Collins, Briggs, and Conway Morris, 1983; and Briggs and Collins, 1988.)

These additional localities are not mere carbon copies of the Burgess. They contain the same basic organisms, but often in very different proportions. One new site, for example, entirely lacks *Marrella*—the most common species by far in Walcott's original quarry. The champion here is *Alalcomenaeus,* one of the rarest creatures, with only two known examples, in the phyllopod bed. Collins also found a few new species. *Sanctaca-*

ris, as already noted, is especially important as the world's first known chelicerate arthropod. Another specimen, a weird wonder, has yet to be described; it is "a spiny animal with hairy legs, of unknown affinities" (Collins, 1985).

Above all, Collins has supplied the most precious themes of diversity and comparison to supplement Walcott's canonical find. His additional localities include five assemblages sufficiently distinct in mix and numbers of species to be called different assemblages. Significantly, these additional sites include four new stratigraphic levels—all close in time to the phyllo-pod bed, to be sure, but still teaching the crucial lesson that the Burgess fauna represents a stable entity, not an unrepeatable moment during an early evolutionary riot of change.

A few basically soft-bodied Burgess species have lightly skeletonized body parts that can fossilize in ordinary circumstances—notably the scle-rites of *Wiwaxia* and the feeding appendages of *Anomalocaris.* These have long been known from distant localities of other times. But a few bits do not make an assemblage. The Burgess fauna, as a more coherent entity, has now been recognized away from British Columbia, in soft-bodied as-semblages in Idaho and Utah (Conway Morris and Robison, 1982, on *Peytoia;* Briggs and Robison, 1984, on *Anomalocaris;* and Conway Morris and Robison, 1986). These contain some forty genera of arthropods, sponges, priapulids, annelids, medusoids, algae, and unknowns. Most have not yet been formally described, but about 75 percent of the genera also occur in the Burgess Shale. Many species once known only for a moment in time, at a dot in space, now have a broad geographic range and an apprecia-ble, stable duration. Writing about the most common Burgess priapulid, Conway Morris and Robison mark the "notable geographic and strati-graphic extensions of a previously unique occurrence. . . . *Ottoia prolifica* has a range through much of the middle Cambrian (?15 million years) during which time it shows minimal morphological changes" (1986, p. 1).

More exciting still has been the recognition of many Burgess elements in older sediments. The Burgess Shale is Middle Cambrian; the famous explosion that originated modern life occurred just before, during the Lower Cambrian. We would dearly like to know whether Burgess disparity was achieved right away, in the heart of the explosion itself.

Even before the most recent discoveries, a few positive hints were al-ready in hand, notably some Burgess-like elements in the Lower Cambrian soft-bodied Kinzers fauna of Pennsylvania, and a suspected weird wonder from Australia, described as an annelid worm in 1979. Then, in 1987, Conway Morris, Peel, Higgins, Soper, and Davis published a preliminary description of an entire Burgess-like fauna from the mid-to-late Lower

Cambrian of north Greenland. The fauna, like the Burgess itself, is dominated by nontrilobite arthropods. The most abundant creature, about a half inch in length, has a semicircular bivalved carapace; the largest, at about six inches, resembles the Burgess soft-bodied trilobite *Tegopelte.* Existing collections are poor, and the area is, as we say in the trade, "difficult of access." But Simon will be visiting next year, and we can expect some new intellectual adventures. In the meantime, he and his colleagues have made the crucial observation, confirming that the Burgess phenomenon occurred during the Cambrian explosion itself: "The extension of stratigraphic ranges of at least some Burgess Shale–like taxa back into the early Cambrian also suggests that they were an integral part of the initial diversification of metazoans" (1987, p. 182).

Last year, my colleague Phil Signor, knowing of my Burgess interests, sent me a spare reprint from a colleague in China (Zhang and Hou, 1985). I could not read the title, but the Latin name of the subject stood out— *Naraoia.* Chinese publications are notorious for poor photography, but the accompanying plate shows an unmistakable two-valved, soft-bodied trilobite. A key Burgess element had been found half a world away. Far more important, Zhang and Hou date this fossil to the *early* part of the Lower Cambrian.

One creature is tantalizing; but we need whole faunas for sound conclusions. I am delighted to report—for it promises to be the most exciting find since Walcott's original discovery itself—that Hou and colleagues have since published six more papers on their new fauna. If the djinn of my previous fable (see page 62) had returned five years ago and offered me a Burgess-style fauna at any other place and time, I could not have made a better choice. The Chinese fauna is half a world away from British Columbia—thus establishing the global nature of the Burgess phenomenon. Even more crucially, the new finds seem well dated to a time *deep* in the Lower Cambrian. Recall the general anatomy of the Cambrian explosion: an initial period, called Tommotian, of skeletonized bits and pieces without trilobites—the "small shelly fauna"; then the main phase of the Cambrian explosion, called Atdabanian, marked by the first appearance of trilobites and other conventional Cambrian creatures. The Chinese fauna comes from the second trilobite zone of the Atdabanian—right in the heart, and near the very beginning, of the main burst of the Cambrian explosion!

Hou and colleagues describe a rich and well-preserved assemblage, including priapulid and annelid worms, several bivalved arthropods, and three new genera with "merostomoid" body form (Hou, 1987a, 1987b, and 1987c; Sun and Hou, 1987a and 1987b; Hou and Sun, 1988).

The Burgess phenomenon, then, goes right back to the beginning of the Cambrian explosion. In a preliminary report, based on admittedly uncertain dating, Dzik and Lendzion (1988) describe a creature like *Anomalocaris* and a soft-bodied trilobite from Eastern European strata *below* the first appearance of ordinary trilobites. We can no longer doubt that Walcott found products of the Cambrian explosion itself in his slightly later strata of British Columbia. Burgess disparity is astounding enough for a time just 30 to 40 million years after the beginning of the Cambrian. But we cannot even view the Burgess range as accumulating steadily during this relatively short period. The main burst occurred well down in the Lower Cambrian—and probably produced the full Burgess range, if the Chinese fauna proves to be as rich as preliminary accounts suggest. The Burgess Shale represents the slightly later period of stabilization for the products of the Cambrian explosion. But what caused the subsequent decimation, and the consequent pattern of modern life, marked by great gaps between islands of extensive diversity within restricted anatomical designs?

THE TWO GREAT PROBLEMS OF THE BURGESS SHALE

The Burgess revision poses two great problems about the history of life. These are symmetrically disposed about the Burgess fauna itself, one before and one after: First, how, especially in the light of our usual views about evolution as a stately phenomenon, could such disparity arise so quickly? And second, if modern life is a product of Burgess decimation, what aspects of anatomy, what attributes of function, what environmental changes, set the pattern of who would win and who would lose? In short, first the origin, second the differential survival and propagation.

In many ways, the first is a juicier problem for evolutionary theory. How in heaven's name could such disparity have arisen in the first place, whatever the later fortunes of its exemplars? But the second problem is the subject of this book, for the decimation of the Burgess fauna raises the fundamental question that I wish to address about the nature of history. My key experiment in replaying the tape of life begins with the Burgess fauna intact and asks whether an independent act of decimation from the same starting point would yield anything like the same groups and the same history that our planet has witnessed since the Burgess maximum in organic disparity. Hence, I shall shamelessly bypass the first problem—but not without presenting a brief summary of possible explanations, if only because one aspect of the potential solution does bear crucially on the second problem of differential fate.

THE ORIGIN OF THE BURGESS FAUNA

Three major kinds of evolutionary explanation are available for the explosion that led to Burgess disparity. The first is conventional, and has been assumed—largely *faute de mieux*—in almost all published discussions. The last two have points in common and represent recent trends in evolutionary thinking. I have little doubt that a full explanation would involve aspects of all three attitudes.

1. *The first filling of the ecological barrel.* In conventional Darwinian theory, the organism proposes and the environment disposes. Organisms provide raw material in the form of genetic variation expressed in morphological differences. Within a population at any one time, these differences are small and—more important for the basic theory—undirected.* Evolutionary *change* (as opposed to mere variation) is produced by forces of natural selection arising from the external environment (both physical conditions and interactions with other organisms). Since organisms supply only raw material, and since this raw material has been judged as nearly always sufficient for all changes occurring at characteristically stately Darwinian rates, environment becomes the motor for regulating the speed and extent of evolutionary alteration. Therefore, according to conventional theory, the maximal rates of the Cambrian explosion must indicate something odd about environments at that time.

When we then inquire about the environmental oddity that could have engendered the Cambrian explosion, an obvious answer leaps at us. The Cambrian explosion was the first filling of the ecological barrel for multicellular life. This was a time of unparalleled opportunity. Nearly anything could find a place. Life was radiating into empty space and could proliferate at logarithmic rates, like a bacterial cell alone on an agar plate. In the bustle and ferment of this unique period, experimentation reigned in a world virtually free of competition for the one and only time.

---

*Biology textbooks often speak of variation as "random." This is not strictly true. Variations are not random in the literal sense of equally likely in all directions; elephants have no genetic variation for wings. But the sense that "random" means to convey is crucial: nothing about genetics predisposes organisms to vary in adaptive directions. If the environment changes to favor smaller organisms, genetic mutation does not begin to produce biased variation toward diminished size. In other words, variation itself supplies no directional component. Natural selection is the cause of evolutionary change; organic variation is raw material only.

In Darwinian theory, competition is the great regulator. Darwin conceived the world in metaphor as a log with ten thousand wedges, representing species, tightly hammered in along its length. A new species can enter this crowded world only by insinuating itself into a crack and popping another wedge out. Thus, diversity is self-regulating. As the Cambrian explosion proceeded, it drove itself to completion by filling the log with wedges. All later change would occur by a slower process of competition and displacement.

This Darwinian perspective also addresses the obvious objection to the model of the empty barrel as the cause of the Cambrian explosion: Life has suffered some astounding mass extinctions since the Cambrian—the Permian debacle may have wiped out 95 percent or more of all marine species—yet the Burgess phenomenon of explosive disparity never occurred again. Life did rediversify quickly after the Permian extinction, but no new phyla arose; the recolonizers of a depleted earth all remained within the strictures of previous anatomical designs. Yet the early Cambrian and post-Permian worlds were crucially different. Five percent may not be a high rate of survivorship, but no mode of life, no basic ecology, was entirely wiped out by the Permian debacle. The log remained populated, even if the wedges had become broader or more widely spaced. To shift metaphors, all the big spheres remained in the barrel, and only the pebbles in the interstices needed a complete recharging. The Cambrian barrel, on the other hand, was flat empty; the log was unscathed, with nary a woodsman's blow nor a lover's knife scratch (see Erwin, Valentine, and Sepkoski, 1987, for an interesting, quantitative development of this general argument).

This conventional view has been assumed in essentially all the Burgess literature—not as an active argument explicitly supported by Burgess evidence, but as the dues that we all properly pay to traditional explanations when we make a side comment on a subject that has not engaged our primary attention. "Less severe competition" has been the watchword of interpretation. Whittington has written, for example:

> Presumably there was abundant food and space in the varied marine environments which were being occupied initially by these new animals, and competition was less severe than in succeeding periods. In these circumstances diverse combinations of characters may have been possible, as new ways of sensing the surroundings, of obtaining food, of moving about, of forming hard parts, and of behavior (e.g. predation and scavenging) were being evolved. Thus may have arisen strange animals, the remains of some of which we see in the Burgess Shale, and which do not fit into our classifications (1981b, p. 82).

Conway Morris has also supported this traditional view. He wrote to me, in response to my defense of unconventional alternatives to follow: "I think that ecological conditions may have been sufficient to account for the observed morphological diversity. . . . Thus, perhaps the Cambrian explosion can be regarded as one huge example of 'ecological release' " (letter of December 18, 1985).

This argument is simply too sensible to dismiss. I haven't the slightest doubt that the "empty ecological barrel" was a major contributor to Burgess disparity, and that such an explosion could never have occurred in a well-filled world. But I don't for a minute believe that external ecology will explain the entire phenomenon. My main defense for this gut feeling relies upon scale. The Cambrian explosion was too big, too different, and too exclusive. I just can't accept that if organisms always have the potential for diversification of this kind—while only the odd ecology of the Lower Cambrian ever permitted its realization—never, not even once, has a new phylum arisen since Burgess times. Yes, the world has not been so empty again, but some local situations have made a decent approach. What about new land risen from the sea? What about island continents when first invaded by new groups? These are not large barrels, but they are at least fair-to-middling bowls. I have to believe that organisms as well as environments were different in Cambrian times, that the explosion and later quiescence owes as much to a change in organic potential as to an altered ecological status.

Ideas about organisms playing such active roles in channeling their own directions of evolutionary change (not merely supplying raw material for the motor of natural selection) have recently grown in popularity, as the strict forms of conventional Darwinism yield their exclusive sway, while retaining their large and proper influence. Evolution is a dialectic of inside and outside, not ecology pushing malleable structure to a set of adaptive positions in a well-oiled world. Two major theories, described in the next two sections, grant a more active role to organic structure.

2. *A directional history for genetic systems.* In the traditional Darwinian view, morphologies have histories that constrain their future, but genetic material does not "age." Differences in rates and patterns of change are responses of an unchanging material substrate (genes and their actions) to variations in environment that reset the pressures of natural selection.

But perhaps genetic systems do "age" in the sense of becoming "less forgiving of major restructuring" (to cite a phrase from J. W. Valentine, who has thought long and deeply about this problem). Perhaps modern organisms could not spawn a rapid array of fundamentally new designs, no matter what the ecological opportunity.

I have no profound suggestions about the potential nature of this genetic "aging," but simply ask that we consider such an alternative. Our exploding knowledge of development and the mechanics of genetic action should provide, within a decade, the facts and ideas to flesh out this conception. Valentine mentions some possibilities. Were Cambrian genomes simpler and more flexible? Has the evolution of multiple copies for many genes, copies that then diverge into a range of related functions, tied up genomes into webs of interaction not easily broken? Did early genes have fewer interactions with others? Did ancient organisms develop with more direct translation of gene to product, permitting such creatures to interchange and alter their parts separately? Most important, do increased complexity and stereotypy of development from egg to adult put a brake upon potential changes of great magnitude? We cannot, for now, go much beyond such crude and preliminary suggestions.

But I can present a good argument against the usual reason for dismissing such ideas in favor of conventional control by external environment. When evolutionists observe that several unrelated lineages react in the same way at the same time, they usually assume that some force external to the genetics of organisms has provoked the common response (for the genetic systems are too unlike, and a similar push from outside seems the only plausible common cause). We have always viewed the creatures that made the Cambrian explosion as unrelated in just this profound way. After all, they include representatives of nearly all modern phyla, and what could be more different, one from the other, than a trilobite, a snail, a brachiopod, and an echinoderm? These morphological designs were as distinct in the Cambrian as they are today, so we assume that the genetic systems were equally unlike—and that the common evolutionary vigor of all groups must therefore record the external push of ecological opportunity.

But this argument assumes the old view of a long, invisible Precambrian history for creatures that evolved skeletons during the Cambrian explosion. The discovery of the Precambrian Ediacara fauna, with the strong possibility that this first multicellular assemblage may not be ancestral to modern groups (see pages 312–13), suggests that all Cambrian animals, despite their disparity of form, may have diverged not long before from a late Precambrian common ancestor. If so—if they had been separate for only a short time—all Cambrian animals may have carried a very similar genetic mechanism by virtue of their strictly limited time of separate life. No ties bind so strongly as the links of inheritance. In other words, the similar response of Cambrian organisms may reflect the *homology* of a genetic system still largely held in common, and still highly flexible, not only the *analogy* of response to a common external push. Of course, life needed the external push of ecological opportunity, but its ability to respond may have marked a shared genetic heritage, now dissipated.

3. *Early diversification and later locking as a property of systems.* My friend Stu Kauffman of the University of Pennsylvania has developed a model to demonstrate that the Burgess pattern of rapid, maximal disparity followed by later decimation is a general property of systems, explicable without a special hypothesis about early relaxed competition or a directional history for genetic material.

Consider the following metaphor. The earthly stage of life is a complex landscape with thousands of peaks, each a different height. The higher the peak, the greater the success—measured as selective value, morphological complexity, or however you choose—of the organisms on it. Sprinkle a few beginning organisms at random onto the peaks of this landscape and allow them to multiply and to change position. Changes can be large or small, but the small shifts do not concern us here, for they only permit organisms to mount higher on their particular peak and do not produce new body plans. The opportunity for new body plans arises with the rarer large jumps. We define large jumps as those that take an organism so far away from its former home that the new landscape is entirely uncorrelated with the old. Long jumps are enormously risky, but yield great reward for rare success. If you land on a peak higher than your previous home, you thrive and diversify; if you land on a lower peak or in a valley, you're gone.

Now we ask, How often does a large jump yield a successful outcome (a new body plan)? Kauffman proves that the probability of success is quite high at first, but drops precipitously and soon reaches an effective zero—just like the history of life. This pattern matches our intuitions. The first few species are placed on the landscape at random. This means that, on average, half the peaks are higher, half lower, than the initial homes. Therefore, the first long jump has a roughly 50 percent chance of success. But now the triumphant species stands on a higher peak—and the percentage of still loftier peaks has decreased. After a few successful jumps, not many higher peaks remain unoccupied, and the probability of being able to move at all drops precipitously. In fact, if long jumps occur fairly often, all the high peaks will be occupied pretty early in the game, and no one has anyplace to go. So the victors dig in and evolve developmental systems so tied to their peaks that they couldn't change even if the opportunity arose later. Thereafter, all they can do is hang tough on their peak or die. It's a difficult world, and many meet the latter fate, not because ecology is a Darwinian log packed tight with wedges, but because even random extinctions leave spaces now inaccessible to everyone.

Kauffman could even quantify the precipitous decline of possibilities for successful jumps. The waiting time to the next higher peak doubles after each successful jump. (Stu told me that a mountain of athletic data shows that when a record is fractured, the average time to the next break doubles.) If your first success needed only two tries on average, your tenth will require more than a thousand. Soon you have effectively no chance of ever getting anywhere better, for geological time may be long, but it is not infinite.

THE DECIMATION OF THE BURGESS FAUNA

We need no more than the descriptive pattern of Burgess disparity and later decimation to impose a major reform upon our traditional view of life. For the new iconography (see figure 3.72) not only alters but thoroughly inverts the conventional cone of increasing diversity. Instead of a narrow beginning and a constantly expanding upward range, multicellular life reaches its maximal scope at the start, while later decimation leaves only a few surviving designs.

But the inverted iconography, however notable, does not have revolutionary impact by itself because it does not exclude the possibility of a fallback to conventionality. Remember what is at stake! Our most precious hope for the history of life, a hope that we would relinquish with greatest reluctance, involves the concepts of progress and predictability. Since the human mind arose so late, and therefore threatens to demand interpretation as an accidental afterthought in a quirky evolutionary play, we are incited to dig in our heels all the harder and to postulate that all previous life followed a sensible order implying the eventual rise of consciousness. The greatest threat lies in a history of numerous possibilities, each sensible in itself after the fact, but each utterly unpredictable at the outset—and with only one (or a very few) roads leading to anything like our exalted state.

Burgess disparity and later decimation is a worst-case nightmare for this hope of inevitable order. If life started with a handful of simple models and then moved upward, any replay from the initial handful would follow the same basic course, however different the details. But if life started with all its models present, and constructed a later history from just a few survivors, then we face a disturbing possibility. Suppose that only a few will prevail, but all have an equal chance. The history of any surviving set is sensible, but each leads to a world thoroughly different from any other. If the human mind is a product of only one such set, then we may not be randomly evolved in the sense of coin flipping, but our origin is the product of massive historical contingency, and we would probably never arise again even if life's tape could be replayed a thousand times.

But we can wake up from this nightmare—with a simple and obvious conventional argument. Granted, massive extinction occurred and only a few original designs survived. But we need not assume that the extinction was a crap shoot. Suppose that survivors prevailed for cause. The early Cambrian was an era of experimentation. Let a bunch of engineers tinker, and most results don't work worth a damn: the Burgess losers were destined for extinction by faulty anatomical construction. The winners were best adapted and assured of survival by their Darwinian edge. What does it matter if the early Cambrian threw up a hundred possibilities, or a thousand? If only half a dozen worked well enough to prevail in a tough world, then these six would form the rootstocks for all later life no matter how many times we replayed the tape.

This idea of survival for cause based on anatomical deftness or complexity—"superior competitive ability" in the jargon—has been the favored explanation, virtually unchallenged, for the reduction of Burgess disparity, and indeed for all episodes of extinction in the history of life. This traditional interpretation is tightly linked with the conventional view for the origin of Burgess disparity as a filling of the empty ecological barrel. An empty barrel is a forgiving place. It contains so much space that even a clap-trap disaster of anatomical design can hunker down in a cranny and hang on without facing competition from the big boys of superior anatomy. But the party is soon over. The barrel fills, and everyone is thrown into the maelstrom of Darwinian competition. In this "war of all against all," the inefficient survivors from gentler times soon make their permanent exit. Only the powerful gladiators win. Thumbs up for good anatomy!

You will read this interpretation in textbooks, in articles of science magazines, even in the *Yoho National Park Highline,* the official newsletter for the home of the Burgess Shale (1987 edition). Under the headline "Yoho's Fossils Have World Significance," we are told: "The first animals moved into the environment devoid of competition. Later, more efficient life forms held sway only to be supplanted again and again as changing conditions and evolution took its course." And when, in 1988, Parks Canada put out the first tourist brochure for its nation's most famous fossils ("Animals of the Burgess Shale"), they wrote that all creatures outside the bounds of modern phyla (the weird wonders of my text) "appear to have been evolutionary dead ends, destined to be replaced by better-adapted or more efficient organisms."

Whittington and colleagues did not, until recently, challenge this comforting view. It makes too much sense. For example, in the summary comments of his monograph on *Wiwaxia,* Conway Morris explicitly linked the two traditional scenarios—barrel filling as a cause of disparity followed by stringent competition as the source of later extinction:

> It may be that diversification is simply a reflection of the availability of an almost empty ecospace with low levels of competition permitting the evolution of a wide variety of bodyplans, only some of which survived in the increasingly competitive environments through geological time (1985, p. 570).

Briggs made the same point for a French popular audience:

> Perhaps this [disparity] is the result of an absence of competition before all the ecological niches of Cambrian seas were filled. Most of these arthropods rapidly became extinct, no doubt because the least well adapted animals were replaced by others that were better adapted (1985, p. 348).*

Whittington also made the almost automatic equation between survival and adaptive superiority:

> The subsequent eliminations among such a plethora of metazoans, and the radiations of the forms that were best adapted, may have resulted in the emergence of what we recognize in retrospect as phyla (1980, p. 146).

Conway Morris and Whittington put the matter most directly in an article for *Scientific American*—probably the best-read source on the Burgess Shale:

> Many Cambrian animals seem to be pioneering experiments by various metazoan groups, destined to be supplanted in due course by organisms that are better adapted. The trend after the Cambrian radiation appears to be the success and the enrichment in the numbers of species of a relatively few groups at the expense of the extinction of many other groups (1979, p. 133).

Words have subtle power. Phrases that we intend as descriptions betray our notions of cause and ultimate meaning. I suspect that Simon and Harry thought they were only delineating a pattern in this passage, but consider the weight of such phrases as "destined to be supplanted" and "at the expense of." Yes, most died and some proliferated. Our earth has always

---

*I retranslate here, hoping not to repeat one of the greatest absurdities I ever encountered—Milton's *Paradise Lost* translated into German as part of the libretto for Haydn's *Creation*, then retranslated as doggerel for a performance in English that could not use Milton's actual words and still retain Haydn's musical values.

worked on the old principle that many are called and few chosen. But the mere pattern of life and death offers no evidence that survivors directly vanquished the losers. The sources of victory are as varied and mysterious as the four phenomena proclaimed so wonderful that we know them not (Proverbs 30:19)—the way of an eagle in the air, the way of a serpent upon a rock, the way of a ship in the midst of the sea, and the way of a man with a maid.

Arguments that propose adaptive superiority as the basis for survival risk the classic error of circular reasoning. Survival is the phenomenon to be explained, not the proof, *ipso facto,* that those who survived were "better adapted" than those who died. This issue has been kicking around the courtyards of Darwinian theory for more than a century. It even has a name—the "tautology argument." Critics claim that our motto "survival of the fittest" is a meaningless tautology because fitness is defined by survival, and the definition of natural selection reduces to an empty "survival of those who survive."

Creationists have even been known to trot out this argument as a supposed disproof of evolution (Bethell, 1976; see my response in Gould, 1977)—as if more than a century of data could come crashing down through a schoolboy error in syllogistic logic. In fact, the supposed problem has an easy resolution, one that Darwin himself recognized and presented. Fitness—in this context, superior adaptation—cannot be defined after the fact by survival, but must be predictable before the challenge by an analysis of form, physiology, or behavior. As Darwin argued, the deer that should run faster and longer (as indicated by an analysis of bones, joints, and muscles) ought to survive better in a world of dangerous predators. Better survival is a prediction to be tested, not a definition of adaptation.

This requirement applies in exactly the same way to the Burgess fauna. If we wish to assert that Burgess extinctions preserved the best designs and eliminated predictable losers, then we cannot use mere survival as evidence for superiority. We must, in principle, be able to identify winners by recognizing their anatomical excellence, or their competitive edge. Ideally, we should be able to "visit" the Burgess fauna in its heyday, while all its elements flourished, and pick out the species destined for success by some definable, structural advantage.

But if we face the Burgess fauna honestly, we must admit that we have no evidence whatsoever—not a shred—that losers in the great decimation were systematically inferior in adaptive design to those that survived. Anyone can invent a plausible story after the fact. For example, *Anomalocaris,* though the largest of Cambrian predators, did not come up a winner. So I could argue that its unique nutcracker jaw, incapable of closing entirely, and probably working by constriction rather than tearing apart of prey, really wasn't as adaptive as a more conventional jaw made of two pieces clamping together. Perhaps. But I must honestly face the counterfactual situation. Suppose that *Anomalocaris* had lived and flourished. Would I not then have been tempted to say, without any additional evidence, that *Anomalocaris* had survived because its unique jaw worked so well? If so, then I have no reason to identify *Anomalocaris* as destined for failure. I only know that this creature died—and so, eventually, do we all.

As the monographic revisions of Burgess genera continued, and as Harry, Derek, and Simon became more adept at reconstructing such unconventional creatures as functioning organisms, their respect grew for the anatomical integrity and efficient feeding and locomotion of the Burgess oddballs. They talked less and less about "primitive" designs, and labored more and more to identify the functional specializations of Burgess animals—see Briggs (1981a) on the tail of *Odaraia*, Conway Morris (1985) on the protective spines of *Wiwaxia*, Whittington and Briggs (1985) on the inferred mode of swimming for *Anomalocaris*. They wrote less about predictable, ill-adapted losers, and began to acknowledge that we do not know why *Sanctacaris* is cousin to a major living group, while *Opabinia* is a memory frozen into stone. The later articles talk more and more about good fortune. Briggs tacked a proviso onto his claim, quoted earlier, about survival due to superior adaptation: ". . . and also, without doubt, because certain species were luckier than others" (1985, p. 348).

All three scientists also begin to emphasize—as a positive note of interest, not an admission of defeat in the struggle to rank Burgess organisms by adaptive worth—the theme that a contemporary observer could not have selected the organisms destined for success. Whittington wrote of *Aysheaia* as a potential cousin to insects, the greatest of all multicellular success stories:

> Looking forward from the Burgess Shale, it would have been difficult to predict which [the survivors] would have been. *Aysheaia*, slow-moving around sponge colonies, hardly would have looked to be the ancestors of those formidable conquerors of the land, myriapods and insects (1980, p. 145).

Conway Morris wrote that "a hypothetical observer in the Cambrian would presumably have had no means of predicting which of the early metazoans were destined for phylogenetic success as established body plans and which were doomed to extinction" (1985, p. 572). He then commented explicitly on the dangers of circular reasoning. Suppose that the jaw of *Wiwaxia* is homologous with the molluscan radula and that the two groups, as closest cousins, represent alternative Burgess possibilities. Since wiwaxiids died and mollusks lived to diversify, one might be tempted to argue that the wiwaxiid molting cycle was less efficient than the continuous accretionary growth of mollusks. But Conway Morris acknowledged that if wiwaxiids had lived and mollusks died, we could have ginned up just as good an argument about the benefits of molting:

> Nevertheless, molting as a mode of growth is widely used in a number of phyla including arthropods and nematodes, these latter two groups being arguably the most successful of all metazoan phyla. In conclusion, if the clock was turned back so metazoan diversification was allowed to rerun across the Precambrian–Cambrian boundary, it seems possible that the successful body plans emerging from this initial burst of evolution may have included wiwaxiids rather than mollusks (1985, p. 572).

Thus, all three architects of the Burgess revision began with the conventional view that winners conquered by dint of superior adaptation, but eventually concluded that we have no evidence at all to link success with predictably better design. On the contrary, all three developed a strong intuition that Burgess observers would not have been able to pick the winners. The Burgess decimation may have been a true lottery, not the predictable outcome of a war between the United States and Grenada or a world series pitting the 1927 New York Yankees against the Hoboken Has-Beens.

We can now fully appreciate the force of so much patient work in documenting the Burgess arthropods. Whittington and colleagues reconstructed some twenty-five basic body plans. Four led to enormously successful groups, including the dominant animals of our world today; all the others died without issue. Yet, except for the trilobites, each surviving group had only one or two representatives in the Burgess. These animals were not marked for success in any known way. They were not more abundant, more efficient, or more flexible than the others. How could a Burgess observer ever have singled out *Sanctacaris*, an animal known from only half a dozen specimens? How, as Whittington argued, could the Burgess handicapper ever have given the nod to *Aysheaia*, a rare and odd creature crawling about on sponges? Why not bet on the sleek and common *Marrella*, with sweeping spines on its head shield? Why not on *Odaraia*, with its subtle and efficient tail flukes? Why not on *Leanchoilia*, with its complex frontal appendage? Why not on sturdy *Sidneyia*, with nothing fancy but everything in order? If we could wind the tape of life back to the Burgess, why should we not have a different set of winners on a replay? Perhaps, this time, all surviving lineages would be locked into a developmental pattern of biramous limbs, well suited for life in the water but not for successful invasion of the land. Perhaps, therefore, this alternative world would have no cockroaches, no mosquitoes, and no black flies—but also no bees and, ultimately, no pretty flowers.

Extend this theme beyond arthropods to the weird wonders of the Burgess. Why not *Opabinia* and *Wiwaxia?* Why not a world of grazing marine herbivores bearing sclerites, not snail shells? Why not *Anomalocaris*, and a world of marine predators with grasping limbs up front and a jaw like a nutcracker? Why not a Steven Spielberg film with a crusty seaman sucked into the cylindrical mouth of a sea monster, and slowly crushed to death by multiple layers of teeth lining a circular mouth and extending well down into the gullet?

We do not know for sure that the Burgess decimation was a lottery. But we have no evidence that the winners enjoyed adaptive superiority, or that a contemporary handicapper could have designated the survivors. All that we have learned from the finest and most detailed anatomical monographs in twentieth-century paleontology portrays the Burgess losers as adequately specialized and eminently capable.

The idea of decimation as a lottery converts the new iconography of the Burgess Shale into a radical view about the pathways of life and the nature of history. I dedicate this book to exploring the consequences of this view. May our poor and improbable species find joy in its new-found fragility and good fortune! Wouldn't anyone with the slightest sense of adventure, or the most weakly flickering respect for intellect, gladly exchange the old cosmic comfort for a look at something so weird and wonderful—yet so real—as *Opabinia?*

CHAPTER IV

Walcott's Vision and the Nature of History

THE BASIS FOR WALCOTT'S ALLEGIANCE TO THE CONE OF DIVERSITY

A BIOGRAPHICAL NOTE

If Charles Doolittle Walcott had been an ordinary man, his shadow would not loom so large over the Burgess Shale and his fundamental error of the shoehorn might merit no more than a footnote. But Walcott was one of the most extraordinary and powerful scientists that America has ever produced. Moreover, his influence rested squarely upon his deeply conservative and traditional perspective upon life and morality. Therefore, if we can grasp the complex reasons for his firm commitment to the Burgess shoehorn, we may win some general insight into the social and conceptual locks upon scientific innovation.

To be sure, Walcott's name is not well known, even to people generally familiar with the history of American science. But his eclipse from public consciousness only reflects our curiously biased view of the history of science, an attitude virtually guaranteed to miscalculate the importance of people in their own time. We value innovation and discovery—quite rightly, of course. Therefore, our genealogy of intellectual progress becomes a chronological list of precursors, people with hot ideas validated by later judgment—even if these scientists enjoyed no influence whatever during their lifetime, and had no palpable impact upon the practice of their profession. For example, we remember Gregor Mendel for the brilliance of his insights, but one can argue that his work scarcely influenced the history of genetics—except ultimately as a beacon and a symbol. His conclusions were ignored in their time, and became influential only when rediscovered by others.

This curiously prospective style of assessment excludes from later consciousness those powerful scientists who in their own time dominated a field, and may have shaped a hundred careers or a thousand concepts, in the service of conventional views later judged incorrect. But how can we grasp science as a social dynamic if we forget these people? How can we sharpen our proper focus upon lonely innovators if we ignore the dominating context of their opposition? Charles Doolittle Walcott is a premier example of such an overlooked man—a great geologist, an indefatigable worker, a noted synthesizer, a central source of power in the social hierarchy of American science, but not, fundamentally, an intellectual innovator.

Walcott's erasure from memory also has another cause, centered upon a paradox. Many scholars, myself among them, loathe administration (while bearing no animus against administrators). This is, of course, a selfish attitude, but life is short and should not be spent wallowing in unhappiness and incompetence—the twin consequences experienced by most scholars who attempt administration. Since scholars write history, skill in management gets short shrift. But where would science be without its institutions? Isolated genius, despite the romantic myths, usually does little by itself.

To make matters worse, great administrators are doubly expunged from history—first, because scholars rarely choose to write about scientific governance; second, because administrative skill breeds invisibility. Bad or dishonest administrators go down in copiously noted shame. The mark of a well-run institution is a smooth flow that appears effortless, nonconstraining, almost automatic. (How many of you know the name of your local bank's president, unless he has been indicted for embezzlement?) Administrators, of course, are well known to their subordinates and beneficiaries—for we must approach the boss to seek those favors of space and money that define the daily business of academia. But a good administrator's name dies with his passage from power.

Charles Doolittle Walcott was a fine geologist, but he was an even greater administrator. During the last two decades of his life, including the entire period of his work with the Burgess Shale, Walcott was the most powerful scientific administrator in America. He not only ran the Smithsonian Institution from 1907 until his death in 1927; he also had his finger—or rather, his fist—in every important scientific pot in Washington. He knew every president from Theodore Roosevelt to Calvin Coolidge, some intimately.* He played a key role in persuading Andrew Carnegie to found the Carnegie Institution of Washington, and worked with Woodrow Wilson to establish the National Research Council. He served as president of the National Academy of Sciences and the American Association for the Advancement of Science. He was a pioneer, booster, and facilitator in the development of American aviation.

Walcott occupied all these roles with grace and consummate skill. Among those who know the Smithsonian's history, I note a virtual consensus in identifying Walcott as the finest secretary between founder Joseph Henry and that recently retired genius of administration, S. Dillon Ripley. Walcott's terse summary at the end of his diary for 1920 provides a good sense of his life at age seventy, at the apex of his power:

> I am now Secretary of Smithsonian Institution, President National Academy of Sciences, Vice Chairman National Research Council, Chairman Executive Committee Carnegie Institute of Washington, Chairman National Advisory Committee for Aeronautics. . . . Too much but it is difficult to get out when once thoroughly immersed in the work of any organization.

Walcott's biography is an American success story. He was born in 1850 and raised near Utica, New York, in a family of barely adequate means. He attended the Utica public schools, but never earned an advanced degree (though numerous honorary doctorates graced his later career). While working on a local farm, he collected trilobites, and took his first step toward a professional career in science by selling his specimens to Louis Agassiz, America's greatest natural historian. (This tale includes a precious irony with respect to later work on the Burgess. Agassiz praised Walcott and bought his collection because Walcott had found trilobite appendages for the first time. Walcott was able to make his discovery because he had recognized the three-dimensional preservation of his fossils, and had noted legs under the carapace. Yet Walcott's principal failure with the Burgess lay in his treatment of these fossils as flat sheets, while Whittington sparked the modern revision by revealing their three-dimensional structure.)

Agassiz's death in 1873 derailed Walcott's hope for formal study in

---

*Perhaps the most touching document in the Walcott archives at the Smithsonian Institution is the highly personal note of condolence written to Walcott by Roosevelt upon the accidental death of Walcott's second wife.

paleontology at Harvard. In 1876, he began his scientific career as assistant to the official New York State geologist, James Hall. He joined the United States Geological Survey in 1879 at the lowest rank of field geologist. By 1894, he had risen to director, firmly guiding the institution through its worst period of financial crisis to a conspicuous rebuilding. He served in this role until his appointment as head of the Smithsonian in 1907.

All this time, Walcott maintained an active and distinguished program of field research and publication on the geology and paleontology of Cambrian strata. He was obsessed with the problem of the Cambrian explosion, and studied Precambrian and Cambrian rocks throughout the world, hoping to achieve some empirical solution. When he found the Burgess Shale in 1909, Walcott was not only the most powerful scientist in Washington but also one of the world's foremost experts on fossil trilobites and Cambrian geology. Charles Doolittle Walcott was no ordinary man.

THE MUNDANE REASON FOR WALCOTT'S FAILURE

As a meticulous and conservative administrator, Walcott left an unintended but priceless gift to future historians. He copied every letter, saved every scrap of correspondence, never missed a day of writing in his diary, and threw nothing out. Even at the very worst moment of his life, when his second wife died in a train crash on July 11, 1911, Walcott wrote a crisply factual entry in his diary: "Helena killed at Bridgeport, Conn. by train being smashed up at 2:30 A.M. Did not hear of it until 3 P.M. Left for Bridgeport 5:35 P.M. . . ." (Walcott may have been meticulous, but please do not think him callous. On July 12, overcome with grief, he wrote: "She was killed by blow on temple (right). . . . I went home where Helena lives in everything about it. My love—my wife—my comrade for 24 years. I thank God that I had her for that time. Her untimely fate I cannot now understand.")

All this material is now housed in eighty-eight large boxes, occupying, as the official report tells us (Massa, 1984, p. 1), "11.51 linear meters of shelf space plus oversize material" in the archives of the Smithsonian Institution. No set of documents can capture the elusive (and mythical) "essence" of a person, for each source tells a piece of the story in yet another way. But the Walcott material is rich and diverse—field notebooks, diaries, private jottings, formal correspondence, business accounts, panoramic photographs, an unpublished "official" biography commissioned by his third wife, tax receipts, diplomas for honorary degrees, letters to his daughter's chaperone and to the custodians of his son's wartime grave in France—and it enables us to construct a revealing picture of this intensely private man who lived in the corridors of public power.

I did not approach the Walcott archives with any general biographical intent. I had but one goal, which became something of an obsession: I wanted to know why Walcott had committed his cardinal error of the shoehorn. I felt that the answer to this question could complete the larger story told by the Burgess Shale—for if Walcott's reasons were rooted not in personal idiosyncrasy, but in his allegiance to traditional attitudes and values, then I could show how Whittington's revision, with its theme of decimation by lottery, overturned something old and central to our culture. I searched through box after box and found numerous clues to a complex set of factors, all clearly indicating that Walcott had been driven to the shoehorn from the core of his being and beliefs. Walcott imposed his well-formulated view of life upon the Burgess fossils; they did not talk back to him in any innovative or independent terms. The shoehorn was a conventional device that preserved both the traditional iconography of the cone of diversity, and its underlying conceptual apparatus of progress and the predictable evolution of consciousness.

My claim may strike many readers as odd and cynical, especially as applied to a scientific theory. Most of us are not naive enough to believe the old myth that scientists are paragons of unprejudiced objectivity, equally open to all possibilities, and reaching conclusions only by the weight of evidence and logic of argument. We understand that biases, preferences, social values, and psychological attitudes all play a strong role in the process of discovery. However, we should not be driven to the opposite extreme of complete cynicism—the view that objective evidence plays no role, that perceptions of truth are entirely relative, and that scientific conclusions are just another form of aesthetic preference. Science, as actually practiced, is a complex dialogue between data and preconceptions. Yet I am arguing that Walcott's shoehorn operated virtually without constraint from Burgess data, and am thus denying that the usual dialogue occurred in this case. Moreover, I make this claim about the greatest discovery of a first-rank scientist, not about a minor episode in the life of a peripheral actor. Can such an unusual one-way flow from preconception to evidence really occur?

Ordinarily, the answer would be no. The fossils would talk back, just as *Opabinia* told Harry Whittington, "I have no legs under my carapace," while *Anomalocaris* exclaimed, "That jellyfish *Peytoia* is really my mouth." But the Burgess animals said little to Walcott, for two basic reasons—thereby casting his shoehorn as a striking example of ideological constraint. First, his preconceptions were strong, rooted as they were at the heart of his social values and the core of his temperament. Second—a reason so ridiculously simple and obvious that we might pass it by in our search for "deeper" meanings—the fossils didn't respond because Walcott never found time to converse with them. A life can be stretched only so far. Administrative burdens did eventually undo Walcott as a working scientist. He simply never found time to study the Burgess specimens. Walcott published four preliminary papers in 1911 and 1912. His associate Charles E. Resser brought out Walcott's posthumous notes in 1931. In between, for the last fifteen years of his busy life, Walcott published monographs on Burgess sponges and algae, but nothing more on the complex animals of the world's most important fossil fauna.

The first reason (strong preconceptions) provides an underpinning for the message of this book; the second reason (administrative burden) is idiosyncratic to Walcott. Yet I begin my discussion with Walcott's idiosyncrasy, for we must understand how he failed to listen before we mount the record of his own song.

Since administrators are usually recruited from the ranks of successful researchers as they reach mid-life, Walcott's story of intensely conflicting demands, and consequent internal stress, echoes a pervasive and honest refrain heard from the helm of scientific institutions. Administrators are chosen because they understand research—meaning that they both love the work and do it well. The story is as old as Walcott's beloved Cambrian mountains. You begin with a promise to yourself: I won't have as much time for research, but I will be more efficient. Others have fallen by the wayside, but I will be different; I will never abandon my research; I will keep working and publishing at close to full volume. Slowly, the perverseness of creeping inevitability takes over. Research fades. You never abandon the ideal, or the original love. You will get back to it, after this term as director, after retirement, after. . . . Sometimes, you really do enjoy an old age of renewed scholarship; more often, as in Walcott's case, death intervenes.

Walcott amazes me. His administrative burdens were so extraordinarily heavy, yet he did continue to publish throughout his later life. His complete bibliography (in Taft* et al., 1928) lists eighty-nine items between 1910, the year of his first report on the Burgess Shale, and 1927, when he died. Fifty-three of these are primary, data-based technical papers. They include major works in taxonomy and anatomy, some written in his busiest years—a hundred pages on Cambrian brachiopods in 1924, eighty on Cambrian trilobites in 1925, a hundred on the anatomy of the trilobite *Neolenus* in 1921. But the Lord's limit of twenty-four hours a day still

*Yes, this is William Howard Taft, then ex-president, and acting chief justice of the United States, who introduced this memorial meeting for Walcott.

grievously restricted Walcott's hopes and plans. Most research did shift to the back burners. The most prominently simmering pot held the fossils of the Burgess Shale. Walcott's guilt at their neglect, and his anticipatory joy in finally returning to his favorite fossils, form a persistent theme in his correspondence. I think that Walcott was consciously saving the Burgess specimens as a primary focus for his years in retirement. But he died with his boots on at seventy-seven.

The whole familiar process, in all its inevitable movement from youthful idealism to elderly resignation, can be traced with unusual thoroughness in the Walcott archives (figures 4.1 and 4.2). On June 2, 1879, the young Walcott, seeking his first job with the U.S. Geological Survey, wrote to the great geologist Clarence King:

> I am willing to do any work that I am able to do that will be of most service. My desire is to pursue stratigraphical geology including collecting and invertebrate paleontology. . . . I desire to make this my life work. . . . I sincerely hope that I may have a trial and then remain or not as my work may decide.

4.1. Charles Doolittle Walcott as a handsome young man of twenty-three. Taken in 1873.

4.2. A photographic portrait of Walcott made about 1915. There are many such portraits in the Smithsonian archives, but I particularly like this one because it seems to show so well both Walcott's strength and great sadness during these years of family tragedy.

King replied positively, and with kindness, on July 18:

> I have given [you] a place at the bottom of the ladder, it will be for you to mount by your own strength. . . . Nothing will give me greater pleasure than to record your work as good.

Walcott's work was better than good, and he rose steadily. By 1893, now near the top of Survey personnel, and firmly committed to a lifetime program of empirical work on older Paleozoic rocks, Walcott refused a teaching job at the University of Chicago in order to continue his research without encumbrance. He expressed his regrets to the preeminent Chicago geologist and administrator T. C. Chamberlin: "As you well know, my desire and ambition is to complete the work on the older Paleozoic formations of the continent and to give to geologists the means of classifying and mapping them."

But in the very next year, 1894, administration called to curtail his work from within. In a letter to his mother, Walcott expressed the conflicting feelings that would haunt him for the rest of his life—pride in recognition, and an urge to serve well, coupled with anxiety about the loss of time for research:

> 10/25/94
> Dear Mother
>   It seems almost strange to me that I am in charge of this great Survey. It is an ever present reality but I have not looked forward to it and still feel the strong desire to resume my old work. I am glad it came to me while you were still with us and I hope that you will live to see the Survey prosper under my administration.
>   With love,
>   Charlie

Thereafter, the theme of conflict between administrative duties and research desires came to dominate Walcott's thoughts. By 1904, while still leading the Geological Survey and before discovering the Burgess, Walcott was already lamenting a massive loss of time for research. On June 18, 1904, he wrote to the geologist R. T. Hill:

> The only personal ambition that I have or have had, that would influence me greatly, is the desire to complete the work on the Cambrian rocks and faunas, which was begun many years ago and which has practically been laid aside for several years past. I hope to give a little time to it this summer, and to do what I can from time to time to complete it. If circumstances were such that I could do it wisely I would most gladly turn over all administration to someone else, and take up my work where I left it in 1892.

Three years later, Walcott assumed his final post, as secretary of the Smithsonian. At the end of this decade, he found the Burgess Shale. Circumstances then conspired, with Walcott's active encouragement, despite his laments, to augment his public responsibilities continuously, and to rob time from any serious or protracted study of the Burgess fossils.

The archives present a panoply of vignettes, glimpses of the multifarious, largely trivial, but always time-consuming daily duties of a chief administrator. He acted on behalf of friends, proposing Herbert Hoover for membership in the American Philosophical Society in 1917. He encouraged colleagues, writing to R. H. Goddard in 1923: "I trust that your work on the 'rocket' is advancing in a satisfactory manner and that in due time you will reach a practical solution of all the problems connected with it." He promoted the welfare of scientists, writing to the chairman of the

Interstate Commerce Commission in 1926 to argue that researchers should receive free railroad passes "in the same category as persons exclusively engaged in charitable or eleemosynary work." He endured endless demands for bits and pieces of his day, as when chief Smithsonian anthropologist Aleš Hrdlička asked for extra time in 1924 to make some forgotten measurements:

> About a year ago when I had the pleasure of measuring you for the records of the National Academy, I did not take the measurement of the hand, foot and a few other parts. Since then, as the result of the analysis of my records on the Old Americans, it has appeared that the dimensions of these parts are of very considerable interest. . . . I should be very grateful if, on an occasion, you would stop in my laboratory for two or three minutes to permit me to take these remaining measurements.

But I found nothing more symbolic, yet so immediately practical, than this affidavit submitted to a bank in 1917 in order to verify a change in his signature: "I enclose herewith the affidavit that you wish. I used to sign my name Chas. D. Walcott. I now use only the initials, as I find it takes too much time to add in the extra letters when there is a large number of papers or letters to be signed."

If these "ordinary" pressures of high administration were not enough to derail research, the decade of 1910 to 1920—spanning his field studies of the Burgess Shale—was full of draining family tragedy for Walcott, as he lost his second wife and two of his three sons (figure 4.3). His son Charles junior died of tuberculosis in 1913, after Walcott had tracked down and evaluated every sanitarium, every rest, dietary, or medical cure, then promoted in the name of hope or quackery. Another son, Stuart, was shot down in an air battle over France in 1917. Walcott wrote to his friend Theodore Roosevelt, who had lost a brother in similar circumstances:

> Stuart, who was in the Western High School in Washington with your brother Quentin, is resting on a hillside in the Ardennes, having been shot down under almost identical circumstance as Quentin, in an air battle with the Huns. He and the two men he brought down are buried at the same place, and a well built cross placed over Stuart's grave bearing his name and the date. When the Huns left they burned and destroyed all the nearby peasant cottages, thus illustrating in the one case their sentimental side and in the other the brute in their nature.

As mentioned previously, Walcott's wife Helena was killed in a train crash in 1911, and his daughter Helen was then sent to Europe, to recover from

4.3. The entire Walcott family in Provo, Utah, in 1907. Standing, from left to right: Sidney, age fifteen; Charles junior, age nineteen; Charles, age fifty-seven; Helena, age forty-two; Stuart, age eleven. Sitting: Helen, age thirteen.

the shock on a grand tour in the company of a chaperone named Anna Horsey. Walcott maintained almost daily contact with the pair, often stepping in to make "appropriate" paternal decisions to guard a beautiful and naive daughter against the perils of impropriety. Walcott's frequent interventions were much appreciated by Ms. Horsey. For example, on June 18, 1912, she wrote: "Your letter has made her realize how objectionable it is for women to smoke. I have told her so often but she thinks I am hopelessly old fashioned." But Ms. Horsey continued to worry. Writing from Paris on July 17, 1912, she warned: "Her beauty is so striking . . . but unless her craving for men's admiration and attention and her extravagant dressing is checked systematically for some time to come, it may lead to great unhappiness." And, in a letter from Italy, she declared: "It truly is not safe. Helen is full of fun and desire for adventure—all girls are at 17—and she is innocent and ignorant and might be induced to meet [men] outside, just for a lark. In Italy, this would be dangerous."

Amidst these extraordinary personal tragedies, the regular affairs of family and business also ate into Walcott's time. He worked with millions invested in the Telluride Power Company, while advising a local bank about the importance of limited credit for his son:

> My son, B. S. Walcott, is a freshman at Princeton. He has an allowance and up to date has been accustomed to paying his bills promptly. I would not, however, credit him or any other boy for more than 30 days, and then only to a limited amount. The effect of credit is bad on the boy and apt to lead to complications.

How could the Burgess Shale possibly have fitted into this caldron, this madhouse of imposed and necessary activity? Walcott needed his summers in the Canadian Rockies for collecting—if only as therapy. But he could never find time for scientific study of the specimens in Washington. A telling indication of Walcott's own growing realization of his predicament may be found in the most revealing set of letters on the Burgess fossils themselves—his correspondence with his former assistant Charles Schuchert, then professor at Yale and one of America's leading paleontologists. In 1912, Walcott was embroiled in committee work, but anticipated only a minor delay in studying some trilobites that Schuchert had sent:

> As to the trilobites, I will not express an opinion until I have a chance to study the whole group next week. I have been so busy with Congressional Committees and other matters the past 10 days that there has been very little opportunity for research.

By 1926, he had admitted defeat, and put off into an indefinite future something far less time-consuming than the study of specimens—the consideration of an argument raised by Schuchert about the anatomy of trilobites: "Someday when I get time I will look over your comments about the structure of trilobites. At present, I am too busy with administrative work."

Several statements from the end of Walcott's life well illustrate his conflicts, his hopes, and the inevitability of his failure to study the Burgess fossils properly. On January 8, 1925, he told the French paleontologist Charles Barrois that he was slowly shedding administrative roles in order to study the Burgess fossils:

> I hope to take up a considerable group of Burgess Shale fossils of great interest, which have not yet been published. Over 100 drawings and photographs have been prepared. They would have been published before this if it

had not been for the time given to administrative duties and matters connected with our scientific organization. I am about through with the latter, as I gave my address as retiring president of the American Association for the Advancement of Science on 12/29, and am also out of the Council of the National Academy. I am planning to resign as a member of the Board of three organizations that are carrying on most interesting and valuable work, but I think my duty to them has been done.

On April 1, 1926, in a letter to L. S. Rowe, Walcott combined his genuine love for research with the canonical, but I think disingenuous, claim that administration was neither enjoyed nor deemed important (relative to scholarship), but only done from sense of duty. (I do not believe that most people are sufficiently self-sacrificial to spend the best years of a life on something that they could put aside with no loss of respect, but only of power. The ethos of science requires that administration be publicly identified as done for duty, but surely most people in such roles take pleasure in their responsibility and influence):

> I would derive the greatest happiness from being able to go on with my research work up to the point of placing on record the data which I have been gathering for the past 15 years in the mountains of the West. . . . Administrative duties have not been unpleasant or disappointing, but I regard them as a passing incident, and not serious work, although of course at times one is called upon to put his best efforts into the solution of the questions that arise.

A week later, he wrote to David Starr Jordan, the great ichthyologist who had served as president of Stanford University, and had been more successful than Walcott in shedding administrative burdens:

> You were a wise man to free yourself from administrative duties. I hope to do so in due time and be free to do some of the things that I have been dreaming of for the past 50 years. It has been a pleasure to dream of them in the past, and every hour that I can get in my laboratory for work is a delight.

On September 27, 1926, Walcott took some action to implement this dream. He wrote to Andrew D. White:

> I wish very much to have a talk with you *in re* Smithsonian Institution and my withdrawing from all executive and administrative work May 1, 1927—when I will have completed 20 years active service as Secretary. Henry, Baird, and Langley died in office but I do not think it is wise for the Smithsonian Institution or for me to go on. I have writing to do that will take all my energy up to 1949. . . . What fun it would be to watch the evolution of democracy up

to 1950. Just now I am not looking ahead beyond 1930. I was told I might pass on at 26, again at 38 and 55 but being of an obdurate temperament I declined.

Charles Doolittle Walcott died in office on February 9, 1927. His remaining, heavily annotated notes on Burgess fossils were published in 1931.

THE DEEPER RATIONALE FOR WALCOTT'S SHOEHORN

Walcott's failure to give his Burgess fossils adequate scrutiny left him free to interpret them along the path of least resistance. Virtually unconstrained by the truly odd anatomy of his specimens, Walcott read the Burgess Shale in the light of his well-established view of life—and the fossils therefore reflected his preconceptions. Since Walcott was such a conservative stalwart—an archtraditionalist not by jerk of the knee but by deep and well-considered conviction—he becomes the finest symbol that I have ever encountered for the embodiment of conventional beliefs.*

To unravel the mystery of the shoehorn, we need to consider Walcott's traditionalism at three levels of increasing specificity—the general cast of his political and social beliefs, his attitude toward organisms and their history, and his approach to the particular problems of the Cambrian.

Walcott's persona

Walcott, an "old American" with rural roots and pure Anglo-Saxon background, became a wealthy man, primarily through judicious investment in power companies. He moved, at least for the last thirty years of his life, in the highest social circles of Washington as an intimate of several presidents and some of America's greatest industrial magnates, including Andrew Carnegie and John D. Rockefeller. He was a conservative by belief, a Republican in politics, and a devout Presbyterian who almost never missed (or failed to record in his diary) a Sunday morning in church.

The letters already quoted have provided some insight into his traditional social attitudes—his differential treatment of sons and daughter, his

---

*I do not like to discuss intellectual issues as abstract generalities. I believe that conceptions are best appreciated and understood through their illustration in a person's idea, or in a natural object. Thus, I am charmed and fascinated by Walcott. I have rarely "met" a man so out of tune with my own view of life—and I do feel that I know him after so much intimacy from the archives. Yet I have gained enormous respect for Walcott's integrity and demoniacal energy in research and administration. I do not particularly like him (as if my opinion mattered a damn), but I am mighty glad that he graced my profession.

ideas on frugality and responsibility. The archives reveal many other facets of this basic personality; I present a small sample just to provide a "feel" for the attitudes of a powerful conservative thinker during the last great age of confidence in American secular might and moral superiority.

In 1923, Walcott wrote to John D. Rockefeller about religion:

> I was brought up at Utica, New York, by my mother and sister, who were consistent Christian women, and I have always adhered to the Presbyterian Church, as I believe in the essentials of the Christian religion and in carrying them out in cooperation with people who believe in the efficacy of the Church as an agency for the preservation and upbuilding of the human race.

I cite Walcott's views on alcohol (to W. P. Eno on October 6, 1923), not because I regard them as quaint or antediluvian (in fact, I agree with Walcott's individual stand, while doubting the political consequences that he envisions in the second paragraph), but because I regard the tone of this passage as so evocative of Walcott's personality and general attitudes:

> When I came to Washington 40 years ago, I used to meet with a group of young men in the afternoon to talk over matters of mutual interest, and we usually had beer and, those who wished, brandy or cocktails. I cared little for any of the drinks and concluded that I was just as well off without them. As time passed on, the homeopathic doses of alcohol gradually showed their effect upon the men by a certain deterioration of character, willpower and effectiveness, and years before they should have done so they passed out [he means died, not collapsed in inebriation] mainly as the result of difficulties with the liver, kidneys and stomach. Only one of them is living today and he gave up "nipping" twenty years ago or more.
>
> I believe that if all alcoholic drinks could be absolutely dispensed with, the betterment and welfare of the human race would be so improved in a generation or two that a large percentage of the suffering, immorality and decadence of individuals and peoples would disappear.

In politics, Walcott seesawed between the conservative poles of jingoism and libertarian respect for untrammeled individual opportunity. In the latter mode, for example, he rejected the labeling of entire races or social classes as biologically inferior, and argued for equal access to education, so that socially widespread genius might always surface. He wrote to Mrs. Russell Sage on June 30, 1913:

> I am particularly interested in your educational work as I believe that it is through education that the great masses of the people are to be brought up to

a standard that will enable them to live healthful, clean lives.

It seems that talent or genius appears about as frequently in one social class as another, in working class children as in the children of the well-to-do. The fact that through the centuries most of the great men have sprung from the comfortable classes simply proves the might of opportunity.

Walcott's jingoistic side emerged particularly in his anger toward Germany over World War I, where he lost a son in aerial combat. In a letter of December 11, 1918, he declined an invitation from the president of Princeton University to a memorial service for students who had died in battle (Walcott frequently used the common epithet of his generation in referring to Germans as Huns):

> I have avoided all memorial meetings and services as the effect upon me is detrimental to my mental and moral poise owing to the depth of feelings aroused against the "Tribe of the Huns" and their allies. This feeling began with the invasion of Belgium, was emphasized by the sinking of the *Lusitania* and the many crimes committed during the war, and now it is not lessened by the many events that have taken place since the signing of the armistice.

All the worst of Walcott's venom poured forth, as the archives reveal, in his confidential spearheading of an extraordinary campaign against the eminent anthropologist Franz Boas in 1920. Boas, as German by birth, Jewish by origin, left-leaning in politics, and pro-German in sympathy, inspired wrath from each and every corner of Walcott's prejudices. In the December 12, 1919, issue of the *Nation,* Boas had published a short letter, entitled "Scientists as Spies," charging that several anthropologists had gathered intelligence data for America during the war while claiming the immunity of science to gain access to areas and information that might otherwise have been declared off limits. He argued that although surreptitious gathering of intelligence is acceptable for men of politics, business, or the military because these professions practice duplicity as a norm, such chicanery can only be viewed as heinous and destructive of scientific principles. Boas's letter would raise few emotions today, and would be read by most people as a somewhat naive evocation of scientific ideals.

But reactions were different in the intensely jingoistic climate of postwar America. To Walcott, Boas's letter was the last straw from a long-standing, disloyal, foreign nuisance. Boas, he claimed, had directly accused President Wilson of lying, for Wilson had stated that "only autocracies maintain spies; these are not needed in democracies." Walcott also interpreted Boas's letter as impugning the integrity of American science *in toto* because a handful of practitioners might have acted as "double agents," both for knowledge and intelligence.

Walcott used this exaggerated reading as the basis for a vigorous campaign to censure Boas, and perhaps to drive him out of American science altogether. Walcott immediately and peremptorily canceled Boas's honorary position at the Smithsonian. He then wrote to all his important and well-placed conservative colleagues, seeking advice on how Boas might be punished. For example, to Nicholas Murray Butler, president of Columbia University (where Boas taught), Walcott wrote on January 3, 1920:

> The position that Dr. Boas had in connection with the Smithsonian Institution was abolished, as it was specially created for him by Secretary Langley in 1901.
>
> The article published by Dr. Boas in the Nation of 12/20 was of such a character that I did not consider a man holding such sentiments a proper person to have an official connection with the Smithsonian. I prefer to have 100 per cent Americans, and have no use personally or officially for the addle-minded Bolshevik type, whether it be Russian or German, Hebrew or Gentile. I realize that the fighting is over with Germany, but it is only begun with the elements that would spread distrust, internal conflict, and ultimate ruin to all that Americans have stood for.

Many colleagues offered the sound advice that if Walcott would simply cool off, the whole matter would soon blow over. Others joined him in McCarthyite frenzy. Writing from Columbia, Michael Pupin longed for the good old days, when men were men and could be mobilized to eliminate such scourges:

> He [Boas] attacks the United States for the purpose of defending Germany, and yet he is allowed to teach our youth and enjoy the honors of being a member of the National Academy of Sciences. This thought makes me long for the good old days of absolutism when the means were always at hand for ridding oneselfe [sic] of such a nuisance as Franz Boas (letter of January 12, 1920).

Walcott heartily agreed: "Thanks for your letter of January 12. It sums up the case of Boas in a very forcible, and to me satisfactory, manner."

At the Anthropological Society of Washington, Walcott spearheaded a resolution castigating Boas, and it passed with only one dissenting vote on December 26, 1919. Four days later, the American Anthropological Association, meeting in Cambridge, Massachusetts, condemned Boas by a vote of 21 to 10, with dissenters labeled as "the Boas group." The resolution included the following interesting prescription as a supposed antidote to Boas's attacks on true democracy:

> It is further respectfully asked, in the name of Americanism as against un-Americanism, that Dr. Boas and also the ten members of the American Anthropological Association, who by voting against the latter resolution thus supporting him in his disloyalty, be excluded from participation in any service respecting which any question of loyalty to the United States Government may properly be raised.

It was a jingoistic age, but then, all times have their extremists, and their keepers of the light.

Walcott's general view of life's history and evolution

Walcott considered himself a follower of Darwin. By most modern readings, such a stated allegiance should imply a strong feeling for quirkiness and opportunism in evolutionary pathways, and a deep conviction that the story of life is about adaptation to changing local environments, not general "progress." But Darwin was a complex man; and the label of his name has been applied to several views of life, some mutually contradictory, and with the preferred focus changing from Darwin's century to our own.

Life was not meant to be free from contradiction or ambiguity. Scholars often err in assuming that their exegesis of a great thinker must yield an utterly consistent text. Great scientists may struggle all their lives over certain issues and never reach a resolution. They may feel the tug of conflicting interpretations and succumb to the attractions of both. Their struggle need not end in consistency.

Darwin waged such a long-standing internal battle over the idea of progress. He found himself in an unresolvable bind. He recognized that his basic theory of evolutionary mechanism—natural selection—makes no statement about progress. Natural selection only explains how organisms alter through time in adaptive response to changes in local environments—"descent with modification," in Darwin's words. Darwin identified this denial of general progress in favor of local adjustment as the most radical feature of his theory. To the American paleontologist (and former inhabitant of my office) Alpheus Hyatt, Darwin wrote on December 4, 1872: "After long reflection, I cannot avoid the conviction that no innate tendency to progressive development exists."

But Darwin was both a critic and a beneficiary of Victorian Britain at the height of imperial expansion and industrial triumph. Progress was the watchword of his surrounding culture, and Darwin could not abjure such a central and attractive notion. Hence, in the midst of tweaking conventional comfort with his radical view of change as local adjustment, Darwin also expressed his acceptance of progress as a theme in life's overall history. He wrote: "The inhabitants of each successive period in the world's history have beaten their predecessors in the race for life, and are, insofar, higher in the scale of nature; and this may account for that vague, yet ill-defined sentiment, felt by many paleontologists, that organization on the whole has progressed" (1859, p. 345).

A kind of unsettled consistency can be forged between these apparently contradictory positions. One can argue that Darwin regarded progress as a cumulative side consequence of a basic causal process operating in other terms at any moment. (Anatomical improvement may be viewed as one pathway toward local adjustment; the local adjustments based on advances in general design may result in increased potential for geological longevity, and progress may emerge by this indirect route.) Critics, myself included, have often suggested such a troubled marriage of Darwin's own conflicting views. Yet I think that the more honorable approach lies simply in acknowledging the genuine contradiction. The idea of progress was too big, too confusing, too central, for such a tidy solution. The logic of theory pulled in one direction, social preconceptions in the other. Darwin felt allegiance to both, and never resolved this dilemma into personal consistency.

Darwin has been a chief scientific saint and guru for more than a century now, and since both views are genuinely part of his thinking, succeeding generations have tended to embrace the side of his thought most in tune with the verities or reforms they wish to support. In our age, so little distant from the "progress" of Hiroshima, and so swamped by the perils of industry and weaponry, we tend to take solace in Darwin's clear view of change as local adaptation and progress as social fiction. But in Walcott's generation, particularly for a man of conspicuous success and strong traditionalist inclinations, Darwin's allegiance to progress as life's pathway became the centerpiece of an evolutionist's credo. Walcott considered himself a Darwinian, expressing by this stated allegiance his strong conviction that natural selection assured the survival of superior organisms and the progressive improvement of life on a predictable pathway to consciousness.

Walcott wrote very little about his general, or "philosophical," approach to the history of life; his published works do not provide the explicit clues that we need to resolve the riddle of his allegiance to the Burgess shoehorn. Fortunately, the archives again provide essential documentation; Walcott preferred to work privately and behind the scenes, but he wrote everything down, in a world innocent of paper shredders and self-dialed transatlantic phone calls.

Amidst his continual emphasis on progress and plan in life's history, I found two especially revealing documents. The first is a heavily annotated typescript for a popular lecture, entitled "Searching for the First Forms of Life," and evidently presented between 1892 and 1894.* Walcott told his audience that Darwin had provided the key to unraveling life's history as "a certain order of progression":

> From the beginning of life on earth there was a connection so close and intimate that, if the entire record could be obtained, a perfect chain of life from the lowest organism to the highest would be established.

Walcott then specified the order revealed by paleontology, in a remarkable passage that embodies the key preconceptions of the shoehorn:

> In early times the Cephalopoda ruled, later on the Crustacea came to the fore, then probably fishes took the lead, but were speedily outpowered by the Saurians. These Land and Sea Reptiles then prevailed until Mammalia appeared upon the scene, since when it doubtless became a struggle for supremacy until Man was created. Then came the age of Invention; at first of flint and bone implements, of bows and arrows and fish-hooks; then of spears and shields, swords and guns, lucifer matches, railways, electric telegraphs.

The entire progressionist credo is rolled up into these few words, but three aspects of the passage stand out for me. First, until the invocation of technology for communication and transportation in the last line, the motive force of progress is entirely martial; animals prevail by dint of force and muscle, humans by the ever more potent instruments of war. Second, Walcott recognized no break between biological and social in his smooth continuum of progressive advance. We mount in an unbroken climb through the ranks of organisms, and continue directly upward with the linear improvement of human technology. Third, Walcott was so committed to progress based on conquest and displacement that he didn't catch the inaccuracy in his own formulation. His chain is not, as implied, a sequence of progressive replacements rooted in superior anatomy (ex-

*Walcott is identified on this manuscript as "of the Geological Survey and Honorary Curator of Paleozoic Fossils in the National Museum." He held the honorary curatorial post from 1892 until he became secretary of the Smithsonian in 1907. I assume that he had not yet been appointed director of the Survey, for he would have been so identified. Since he became director in 1894, the date of the lecture must be between 1892 and 1894.

pressed as weaponry) on an eternal battleground. Reptiles did not replace fishes; rather, they represent an oddly modified group of fishes in a novel terrestrial environment. Fishes have never been replaced as dominant vertebrates of the oceans. But Walcott is so committed to an equation between the linear scale of progress by battle and the conventional order of vertebrate taxonomy that he overlooks this basic flaw.

How could such a view of life as a single progressive chain, based on replacement by conquest and extending smoothly from the succession of organic designs through the sequence of human technologies, possibly accommodate anything like our modern interpretation of the Burgess fauna? For Walcott, the Burgess, as old, had to include a limited range of simple precursors for later improved descendants. The modern themes of maximal disparity and decimation by lottery are more than just unacceptable under such a view of life; they are literally incomprehensible. They could never even arise for consideration. For Walcott, the Burgess organisms had to be simple, limited in scope, and ancestral—in other words, products of the conceptual shoehorn. And lest you doubt that Walcott made this logical inference from his own preconceptions, another passage in the same address explicitly restricts all past diversity within the boundaries of a few major lineages, destined for progress: "Nearly all animals, whether living or extinct, are classed under a few primary divisions or morphologic types."*

---

*One tangential point before leaving this rare example of a public address by such a private and imperious man. Walcott was a clear but uninspired writer. So many professionals make the mistake of assuming that popular presentations of science—particularly writing about nature—must abandon clarity for overblown, rapturous description. A Wordsworth or a Thoreau can pull it off; the great majority of naturalists, however great their emotional love for the outdoors, cannot—and should not try, lest the ultimate in unintended parody arise. Besides, audiences do not need such a crutch. The "intelligent layperson" exists in abundance and need not be coddled. Nature shines by herself. But, in any case, and with some embarrassment, I give you Charles Doolittle Walcott on the Grand Canyon at sunset:

> The Western sky is all aflame. The scattered banks of clouds and wavy cirrus have caught the warring splendor, and shine with orange and crimson. Broad slant beams of yellow light, shot through the glory-rifts, fall on turret and tower, on pinnacled crest and wending ledge, suffusing through with a radiance less fulsome, but akin to that which flames in western clouds. The summit band is brilliant yellow, the next below is pale rose. But the grand expanse within is deep, luminous, resplendid [sic] red. The climax has now come; the blaze of sunlight poured over an illimitable surface of glowing red is flung back into the gulf, and, commingling with the blue haze, turns it into a sea of purple of most imperial hue. However vast the magnitudes, however majestic the forms or sumptuous the decoration, it is in these kingly colors that the highest glory of the Grand Canyon is revealed.

If this document were not enough, the second adds a moral and religious dimension to Walcott's need for progress and the Burgess shoehorn. Walcott's simple description of evolutionary pathways was sufficient, by itself, to guarantee the shoehorn and preclude any thought of decimation by lottery. But if you believe that nature also embodies moral principles, and that stately progress and predictability form a basis for ethics, then the internal necessity for the shoehorn increases immeasurably. Description is powerful enough by itself; prescription can overwhelm. On January 7, 1926, Walcott wrote to R. B. Fosdick about the moral value of orderly progress in evolution:

> I have felt for several years that there was danger of science running away with the orderly progress of human evolution and bringing about a catastrophe unless there was some method found of developing to a greater degree the altruistic or, as some would put it, the spiritual nature of man.

The second document on morality and the shoehorn represents Walcott's deeply felt response to a key episode in twentieth-century American social history—the fundamentalist anti-evolution crusade that culminated in the Scopes trial of 1925. Led by the aged but still potent William Jennings Bryan—America's greatest orator and a three-time loser for the presidency (see Gould, 1987c)—biblical literalists had persuaded several state legislatures to ban the teaching of evolution in public schools.

The canonical attitude of scientists then and now—and the argument that finally secured our legal victory before the Supreme Court in 1987—holds that science and religion operate in equally legitimate but separate areas. This "separationist" claim allots the mechanisms and phenomena of nature to scientists and the basis for ethical decisions to theologians and humanists in general—the age of rocks versus the rock of ages, or "how heaven goes" versus "how to go to heaven" in the old one-liners. In exchange for freedom to follow nature down all her pathways, scientists relinquish the temptation to base moral inferences and pronouncements upon the physical state of the world—an excellent and proper arrangement, since the facts of nature embody no moral claims in any case.

To Walcott, this separationist view was anathema. He longed to find moral answers directly in nature—his kind of answers, to support his conservative view of life and society. He wished to bring science and religion together, not carve out separate domains in mutual respect. In fact, he charged that the separationist argument had fanned Bryan's anti-intellectual flame by driving people to the suspicion that scientists really wanted to dispense with religion entirely (but settled, as a temporary and practical matter, for the banning of religion from the affairs of nature). Walcott therefore decided to combat Bryan and his ilk by publishing a statement, signed by a group of respected traditionalists like himself, on the connections between science and religion—particularly on the manifestation of God's handiwork in the pathways of evolutionary change. Canvassing for signatures, he circulated a letter among his friends:

> Unfortunately through the action of radicals in science and in religion, men of the type of mind of William Jennings Bryan have seen a great danger coming to religion through the teaching of the facts of evolution.
>
> A number of conservative scientific men and clergymen have been asked to sign a statement to be given much publicity, on the relations of science and religion.

The statement, published in 1923, two years before the Scopes trial, bore Walcott's name as first signer, and included Herbert Hoover and such scientific leaders as Henry Fairfield Osborn, Edwin Grant Conklin, R. A. Millikan, and Michael Pupin. "In recent controversies," the statement held, "there has been a tendency to present science and religion as irreconcilable and antagonistic domains of thought. . . . They supplement rather than displace or oppose each other."

Walcott's statement went on to argue that the fundamentalist assault could only be quelled by showing the unity of science with religious truths that most Americans viewed as basic to their personal equanimity and social fabric. The primary evidence for this unity lay in the ordered, predictable, and progressive character of life's history—for the pathways of evolution displayed God's continuous benevolence and care for his creation. Evolution, with its principle of natural selection leading to progress, represented God's way of showing himself through nature:

> It is a sublime conception of God which is furnished by science, and one wholly consonant with the highest ideals of religion, when it represents Him as revealing Himself through countless ages in the development of the earth as an abode for man and in the age-long inbreathing of life into its constituent matter, culminating in man with his spiritual nature and all his God-like power.

In this key passage, the shoehorn becomes an instrument of God. If the history of life shows God's direct benevolence in its ordered march to human consciousness, then decimation by lottery, with a hundred thousand possible outcomes (and so very few leading to any species with self-conscious intelligence), cannot be an option for the fossil record. The creatures of the Burgess Shale must be primitive ancestors to an improved set of descendants. The Burgess shoehorn was more than a buttress to a comfortable and convenient view of life; it was also a moral weapon, and virtually a decree of God.

The Burgess shoehorn and Walcott's struggle with the Cambrian explosion

If Walcott had never encountered a Cambrian rock before discovering the Burgess Shale, his persona and general attitude toward evolution would by themselves have generated the shoehorn. But Walcott also had highly specific reasons for his view, based upon his lifelong commitment to Cambrian studies, particularly his obsession with the problem of the Cambrian explosion.

I devoted the first chapter of this book to documenting the influence of iconography upon concepts. I showed how two basic pictures—the ladder of progress and the cone of increasing diversity—buttressed a general view of life based on human hopes, and forced a specific interpretation of Burgess animals as primitive precursors. In the present chapter, my two previous sections, on Walcott's persona and attitude toward evolution, invoke the ladder; his more specific argument about the Cambrian rests upon the cone.

Evolutionary trees as the standard iconography for phylogeny had been introduced in the 1860s by the German morphologist Ernst Haeckel. (Others, including Darwin in his single drawing for the *Origin of Species*, had used botanical metaphors and drawn abstract, branching diagrams as general guides to relationships among organisms. But Haeckel developed this iconography as the preferred representation of evolution. He drew numerous trees with real bark and gnarled branches. And he placed an actual organism on each twig of his copious arborescences.) To native speakers of English, Haeckel's name may not be so well known as Thomas Henry Huxley's, but he was surely the most dogged and influential publicist that ever spoke for evolution. Those trees, the mainstay of instruction when Walcott studied and taught paleontology, embody the themes of ladder and cone in both flamboyantly overt and deceptively subtle ways.

To begin, all of Haeckel's trees branch continually upward and outward, forming a cone (Haeckel sometimes allowed the two peripheral branches in each subcone to grow inward at the top, in order to provide enough room on the page for all groups—but note how he carefully preserved the general impression of up and out whenever he used this device). Haeckel's placement of groups reinforces the great conflation of low with primitive, thus uniting the central themes of cone and ladder.

Consider, for example, Haeckel's treatment of vertebrate phylogeny (figure 4.4; all figures from Haeckel appear in his *Generelle Morphologie* of 1866). The entire tree branches upward and outward, forming two levels, with greater diversity at the top. The lower tier, for fishes and amphibia, clearly denotes limited spread and primitivity; the upper, for reptiles, birds, and mammals, implies both more and better. Yet fishes and amphibians live still, whatever their time of origin—and fishes are by far the most diverse of vertebrates both in range of morphology and number of species. Haeckel's tree of mammals (figure 4.5) dramatically illustrates the conflation of high with advanced, and the misrepresentation of relative diversity that may arise when a small twig is equated with an entire upper level of progress. On this tree, the highly diverse and morphologically specialized artiodactyls (cattle, sheep, deer, giraffes, and their relatives) are squeezed

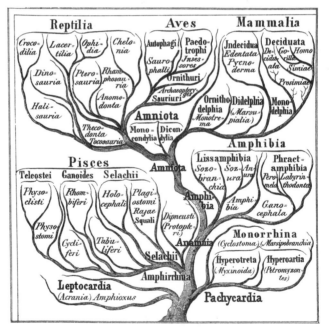

4.4. Haeckel's evolutionary tree of the vertebrates (1866). Fishes (Pisces) actually encompass more disparity than all the rest of the vertebrates combined, but this false iconography, based on the cone of increasing diversity, confines them to a lower branch that gains in breadth as it expands upward.

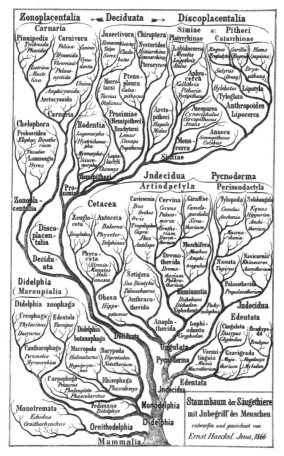

4.5. The evolutionary tree of mammals according to Haeckel (1866).

together in the lower tier. By contrast, the primates, forming a comparatively small group, occupy nearly half the upper level on the culturally favored right-hand side. The most diverse of all mammals, the rodents, must squash into a little bubble of space, caught in limbo between the two main layers—for there is no room for them to spread out at the top, where Haeckel's two favored groups—carnivores (for general valor) and primates (for smarts)—hog all the space.

Echinoderms provide the test case for the iconography of the tree, for in well-preserved hard parts already well-documented in Haeckel's time, they tell the same tale as the Burgess Shale—maximal early disparity followed by decimation. Note how Haeckel acknowledges this maximal early disparity with a forest of primary stems at the geological beginning (figure 4.6). But the cone decrees that trees must spread outward as they grow, so all these early groups are shrunk into the insignificant space available at the outset. The radically decimated modern tree concentrates nearly all its diversity in two groups of strictly limited range in design—the starfish (Haeckel's "Asterida") and the sea urchins (his "Echinida"). Yet Haeckel's iconography conveys the impression of a continuous increase in range.

Finally, consider Haeckel's tree of annelids and arthropods (figure 4.7), the framework upon which Walcott would hang all the Burgess organisms

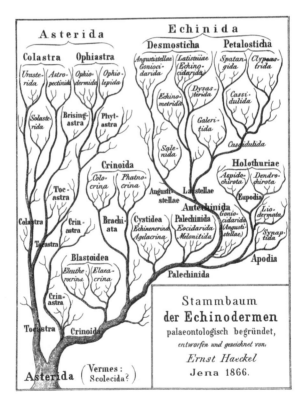

4.6. The evolutionary tree of echinoderms as depicted by Haeckel (1866), in accordance with the cone of increasing diversity. This group actually displays the Burgess pattern of maximal early disparity followed by decimation, but Haeckel's iconography conveys the impression of continuously increasing diversity and range.

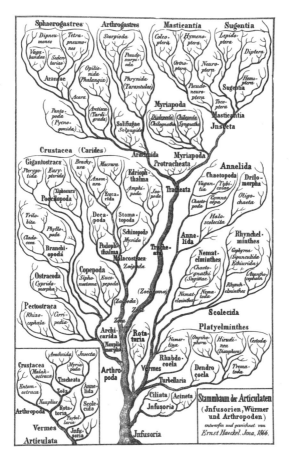

4.7. The evolutionary tree of arthropods and their relatives as depicted by Haeckel (1866), once again in accordance with the cone of increasing diversity.

that have fueled our new interpretation. Upon this ultimate expression of up and out, Walcott put all the Burgess arthropods on two adjacent branches of the lower tier—*Sidneyia* and its relatives in Haeckel's "Poecilopoda" with horseshoe crabs and eurypterids, and nearly all other forms on the branchiopod-trilobite branch.

Walcott followed all these iconographical conventions in the three sketchy trees that represent his only published attempts to draw a phylogeny for Burgess organisms. All appear in his major paper on Burgess arthropods (Walcott, 1912). Considered in their original order, they beautifully illustrate the restriction of ideology by iconography. His first chart (figure 4.8) claims to be a simple description of "stratigraphic distribution" in a

STRATIGRAPHIC DISTRIBUTION OF EARLIEST REPRESENTATIVES OF EACH OF THE FIVE SUBCLASSES OF THE CRUSTACEA.

4.8. Walcott's first chart showing the phylogeny of Burgess arthropods (1912). Walcott forcibly shaped his data in accordance with the cone and ladder by drawing speculative lines of convergence toward a common ancestry in his hypothetical Lipalian interval. He also minimized the explosion of disparity in the Burgess itself by lining up, in an apparently temporal sequence, five forms that were actually contemporaneous (right) and by drawing a hypothetical line at the left boundary to suggest continuing diversity after the Burgess where no evidence exists.

phylogenetic context. Yet even here, both conventions of cone and ladder conspire to confine Burgess disparity within the limits of a few recognized major groups. The *ladder* acts to compress one group of five "merostom-oid" genera into a single line: by treating *Habelia–Molaria–Emeraldella–Amiella–Sidneyia* as a structural sequence of ancestors for eurypterids and horseshoe crabs, Walcott conveyed an impression of temporal succession for these contemporaneous (and, we now know, quite unrelated) genera.

The *cone* then forces all other genera into two major groups—the branchiopod and the trilobite-to-merostome lineages. All these genera were contemporaneous, but Walcott framed the entire picture with two vertical lines, implying that later ranges continued to match recorded Burgess disparity—although no direct evidence supports this assumption. Note, especially, that the left-hand boundary line corresponds to no organism at all—the line is an iconographical device added to guide the eye into seeing a cone. Without this line, disparity would be maximal in the Burgess, and markedly decreased thereafter. Never doubt the power of such tiny and apparently insignificant moves. In a way, everything that I am trying to say in this book achieves an elegant epitome in this one vertical stroke—added to represent a philosophy of life, not the empirical record of organisms.

As a second device, buttressed by no data and added to support a traditional interpretation, Walcott drew the origin of Burgess genera at different levels within a Precambrian interval that he called Lipalian. He connected these levels with two slanted lines that point downward toward a distant Precambrian ancestor for the entire tree. This device provides the tree with a root, in an early period of restricted disparity. But Walcott had no evidence at all—and we have none today—for such evolutionary order among the Burgess arthropods.

Walcott's second chart (figure 4.9) illustrates the tyranny of the cone in an even more striking manner. Walcott claimed that five distinct lineages could be recognized among Burgess arthropods—the extinct trilobites, and four prominent groups of organisms inhabiting modern waters. Again, he used two devices to compress Burgess disparity into the narrow end of a cone. First, he showed all five lineages as converging toward the bottom (subtly for four, perhaps because he felt sheepish about making such an assertion with no supporting data at all; more boldly, with a distinct angular bend, for the merostome lineage, where he adduced some evidence—see below). Second, he placed all these contemporaneous fossils at different positions on his vertical branches, implying that they represented evolutionary diversification through time. On the merostome branch, he lined up eight genera (five of which are known only as contemporaries in the Burgess Shale) to forge a hypothetical link between merostomes and crustaceans: "Such forms as *Habelia, Molaria* and *Emeraldella* serve to fill in the gap between the Branchiopoda and the Merostomata as represented by *Sidneyia* and later the eurypterids" (1912, p. 163). Finally, figure 4.10 shows Walcott's last and most abstract phylogeny for the Burgess arthropods. Even larger groups are lined up on vertical branches, and the entire tree converges to a branchiopod root.

These phylogenies embody the crucial link between Walcott's interpretation of Burgess arthropods and the previous focus of a career that had spanned more than thirty intense years—the study of Cambrian rocks and the problem of the Cambrian explosion. The linkage between the Burgess and Walcott's view of the Cambrian explosion provides a final, and more specific, explanation for his inevitable embrace of the shoehorn as an interpretation for Burgess fossils.

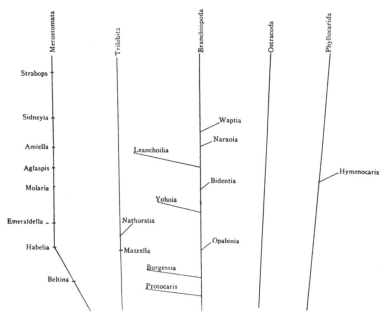

THEORETICAL LINES OF DESCENT OF CAMBRIAN CRUSTACEA

4.9. Walcott's second chart showing the phylogeny of Burgess arthropods (1912). Again, the lineages converge toward a hypothetical common ancestry, and several contemporaneous forms are placed in ladder-like order, on the left-hand and middle lines.

In short, Walcott viewed the Burgess arthropods as members of five major lineages, already stable and well established at this early Cambrian date. But if life had already become so well differentiated along essentially modern lines, the five lineages must have existed at the inception of the Cambrian explosion as recorded by fossil evidence—for evolution is stately and gradual, not a domain of sudden jumps and mad eruptions of diversity. And if the five lineages existed as well-differentiated groups right at the beginning of the Cambrian, then their common ancestor must be sought *far back* in the Precambrian. The Cambrian explosion must therefore be an artifact of an imperfect fossil record; the late Precambrian seas, in Darwin's words, must have "swarmed with living creatures" (1859, p. 307).

Walcott thought that he had discovered why we have no evidence for this necessary Precambrian richness. In other words, he thought that he had solved the riddle of the Cambrian explosion in orthodox Darwinian

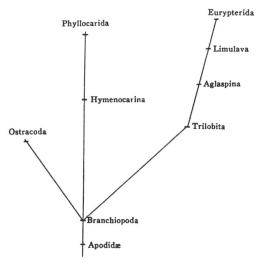

THEORETICAL EVOLUTION OF CAMBRIAN CRUSTACEA FROM THE BRANCHIOPODA

4.10. Walcott's third and last attempt at depicting arthropod evolution (1912). The lineages now converge to a common point, and major groups are lined up, one above the other, on one of the three diverging branches.

terms. The ordering of Burgess arthropods into five well-known and stable groups cemented his solution:

> The Cambrian crustacean fauna suggests that five main lines or stems . . . were in existence at the beginning of Cambrian time and that all of them had already had their inception in Lipalian time or the period of the Precambrian marine sedimentation of which no known part is present in on the existing continents (1912, pp. 160–61).

We must remember that the Cambrian explosion was no ordinary riddle, and its potential solution therefore no minor plum, but something more akin to the Holy Grail. Darwin, as already noted, had publicly fretted that "the case at present must remain inexplicable; and may be truly urged as a valid argument against the views here entertained" (1859, p. 308).

Two different kinds of explanations for the absence of Precambrian ancestors have been debated for more than a century: the artifact theory (they did exist, but the fossil record hasn't preserved them), and the fast-transition theory (they really didn't exist, at least as complex invertebrates easily linked to their descendants, and the evolution of modern anatomical plans occurred with a rapidity that threatens our usual ideas about the stately pace of evolutionary change).

Darwin, making his characteristic (and invalid) conflation of leisurely, gradual evolution and change by natural selection, rejected the fast-transition theory out of hand. He insisted that any complex Cambrian creature must have arisen from a lengthy series of Precambrian ancestors *with the same basic anatomy:* "I cannot doubt that all the Silurian [Cambrian, in modern terminology] trilobites have descended from some one crustacean, which must have lived long before the Silurian [Cambrian] age" (1859, p. 306).

Accordingly, Darwin searched for a believable version of the artifact theory, finally proposing that, in Precambrian times, "clear and open oceans may have existed where our continents now stand." Such tracts of uninterrupted water would have received little or no sediment. Hence our current continents, containing all rocks available to our view, rose from an area that accumulated no strata during the crucial span of late Precambrian faunas, while regions of shallow water that did receive Precambrian sediments now lie in inaccessible oceanic depths.

Walcott had long maintained a firm commitment to the artifact theory. It provided the keystone for his entire approach to Cambrian geology and life. He never doubted that Cambrian complexity and diversity required a long series of abundant Precambrian ancestors of similar anatomy. In an early article he wrote: "That the life in the pre-*Olenellus* sea was large and varied there can be little, if any, doubt. . . . It is only a question of search and favorable conditions to discover it" (1891). *Olenellus,* as then defined, was the oldest Cambrian trilobite, so pre-*Olenellus* meant Precambrian. And in one of his late papers: "When the advanced stage of development of some of the earliest-known forms is considered it seems almost certain that such existed far back in Precambrian time" (1916, p. 249).

Walcott had long defended a particular approach to the artifact theory that a profusion of new Burgess phyla would have undermined. The artifact theory demanded long Precambrian histories for many modern groups, yet no fossils had been found. Therefore, the existence of Precambrian life would have to be inferred from some aspect of later, recorded history. Accordingly, Walcott sought support for the artifact theory in the concept of stability. If the number of basic anatomical designs had not changed throughout life's recorded history, then such stability must surely guide our concept of what came before. Could a system so constant for hundreds of millions of years arise in a geological flash just a moment before? Protracted stability surely implied a very long and stately approach from a common ancestry deep in the distant Precambrian mists, not a gigantic burp of creativity from a starting point just below the Cambrian borderline.

We can now understand why Walcott was virtually compelled to propose the Burgess shoehorn. He interpreted his new fauna in the light of thirty previous years spent (largely in frustration) trying to prove the artifact theory, as an ultimate tribute to Darwin from a Cambrian geologist. He could not grant Burgess organisms the uniqueness that seems so evident to us today because a raft of new phyla would have threatened his most cherished belief. If evolution could produce ten new Cambrian phyla and then wipe them out just as quickly, then what about the surviving Cambrian groups? Why should they have had a long and honorable Precambrian pedigree? Why should they not have originated just before the Cambrian, as the fossil record, read literally, seems to indicate, and as the fast-transition theory proposes? This argument, of course, is a death knell for the artifact theory.

If, instead, he could shoehorn all Burgess creatures into modern groups, he would be giving the strongest possible boost to the artifact theory. For such a condensation of disparity increased the proportion of modern groups already represented right at the start of life's recorded history—and greatly enhanced the apparent stability of major designs through time. Obviously, and with both vigor and delight, Walcott chose this alternative. What does any man do when faced with destruction or affirmation?

Walcott approached the artifact theory from both geological directions—down from the Cambrian, as illustrated by the Burgess shoehorn, and up from the Precambrian. His argument about the Precambrian has become, in the typically perverse manner of textbook histories, his most enduring legacy. Most textbooks contain a traditional, almost mandatory, two- or three-page introductory section on the history of their discipline. These travesties of scholarship dismiss the fine thinkers of our past with two-liners about some error, usually misinterpreted, that shows how stupid they were and how enlightened we have become. Charles Doolittle Walcott was one of the most powerful men in the history of American science. Yet ask any student of geology about him, and if you get any response at all, you will probably hear: "Oh yeah, that doofus who invented the nonexistent Lipalian interval to explain the Cambrian explosion." I first heard of Walcott in this context, long before I knew about the Burgess Shale. History can be either enlightening or cruel. However, with the preceding discussion of the Burgess and the artifact theory in mind, I think we can finally understand the story of the Lipalian interval properly—and recognize Walcott's proposal as a reasonable, if outstandingly wrong, inference within his general commitments.

The artifact theory was central to Walcott's scientific perspective. His conclusions about the Burgess fauna supported this theory, but he needed a more direct argument from the Precambrian. Where had all the Precambrian animals gone? Other ideas included a universal metamorphism (alteration of rocks by heat and pressure) that had destroyed all Precambrian fossils, and an absence of fossilizable hard parts in Precambrian creatures. Walcott rejected the metamorphism theory because he had found many unaltered Precambrian rocks, and he argued that the hard-part theory, while probably true, could not explain the entire phenomenon.

Walcott was primarily a field geologist, specializing in Cambrian rocks. Following the proclivities of any field man, he approached his growing interest in the problem of the Cambrian explosion in the obvious way—he decided to search the latest Precambrian rocks for the elusive ancestors of Cambrian fossils. He worked for many years in the western United States, the Canadian Rockies (where he discovered the Burgess), and in China, but he found no Precambrian fossils. So he tried to reconstruct the geological and topographic history of the late Precambrian earth in a way that would explain this frustrating absence.

Walcott eventually reached a conclusion opposite to Darwin's speculation but in the same tradition—the rocks that might house abundant Precambrian fossils just aren't accessible to us. Darwin had suggested vast Precambrian oceans with no continents nearby to serve as a source of sediments. Walcott argued that the late Precambrian was a time of uplift and mountain building, with continents far more extensive than today's. Since life, according to Walcott and others, had evolved in the oceans and had not yet colonized land or fresh waters, these vast Precambrian continents permitted no marine sedimentation in areas now accessible to us. (Walcott wrote long before the era of continental drift and never doubted the permanent position of continents. Thus, he argued that places available for geological observation today were the centers of more extensive Precambrian continents, and were therefore devoid of late Precambrian marine sediments. Late Precambrian sediments might lie under miles of deep ocean, but no technology then existed to recover or even to sample such potential treasures.)

The infamous "Lipalian interval" was Walcott's name for this time of Precambrian nondeposition. Walcott proposed a world-wide break in accessible marine sedimentation, just during the critical interval of extensive Precambrian ancestry for modern groups. In a famous address to the Eleventh International Geological Congress, meeting in Stockholm on August 18, 1910, he stated:

> I have for the past 18 years watched the geological and paleontological evidence that might aid in solving the problem of Precambrian life. The great series of Cambrian and Precambrian strata in eastern North America from Alabama to Labrador; in western North America from Nevada and California far into Alberta and British Columbia, and also in China, have been studied and searched for evidences of life until the conclusion had gradually been forced upon one that on the North American continent we have no known Precambrian *marine* deposits containing traces of organic remains, and that the abrupt appearance of the Cambrian fauna results from geological and not from biotic conditions. . . . In a word, the thought is that the Algonkian [late Precambrian] period . . . was a period of continental elevation and largely terrigenous sedimentation in non-marine bodies of water, also a period of deposition by aerial and stream processes over considerable areas (1910, pp. 2–4).

And he added:

> Lipalian is proposed for the era of unknown marine sedimentation. . . . The apparently abrupt appearance of the lower Cambrian fauna is . . . to be explained by the absence near our present land area of the sediments, and hence the faunas of the Lipalian period (1910, p. 14).

Walcott's explanation may sound forced and *ad hoc*. It was surely born of frustration, rather than the pleasure of discovery. Yet the nonexistent Lipalian was not a fool's rationalization, as usually presented in our textbooks, but a credible synthesis of geological evidence in the context of a vexatious dilemma. If Walcott deserves any brickbats, direct them at his failure to consider any alternative to his favored way of thinking about the artifact theory—and at his false assumption, imposed by the old bias of gradualism, that equated evolution itself with a long sequence of ancestral continuity for any complex creature. For even if the Lipalian hypothesis made sense in the light of existing geological information, it rested, as Walcott knew only too well, upon the most treacherous kind of argument that a scientist can ever use—negative evidence. Walcott admitted: "I fully realize that the conclusions above outlined are based primarily on the absence of a marine fauna in Algonkian rocks" (1910, p. 6).

And, as so often happens in the face of negative evidence, the earth eventually responded, offering to later geologists abundant late Precambrian marine sediments—still with no fossils of complex invertebrates. The

Lipalian interval ended up on the trash heap of history.

Scientists have a favorite term for describing a phenomenon like Walcott's allegiance to the Burgess shoehorn—overdetermined. The modern concept of maximal disparity and later decimation (perhaps by lottery) never had the ghost of a chance with Walcott because so many elements of his life and soul conspired to guarantee the opposite view of the shoehorn. Any one of these elements would have been enough in itself; together, they overwhelmed any alternative, and overdetermined Walcott's interpretation of his greatest discovery.

To begin, as we have seen, Walcott's persona as an archtraditionalist in thought and practice did not lead him to favor unconventional interpretations in any area of life. His general attitude to life's history and evolution implied stately unfolding along predictable pathways defined by the ladder of progress and cone of increasing diversity; this pattern also held moral meaning, as a display of God's intention to imbue life with consciousness after a long history of upward striving. Walcott's specific approach to the key problem that had focused his entire career—the riddle of the Cambrian explosion—favored a small set of stable and well-separated groups during Burgess times, so that a long history of Precambrian life might be affirmed, and the artifact theory of the Cambrian explosion supported. Finally, if Walcott had been at all inclined to abandon his ideological commitment to the shoehorn, in the light of contradictory data from the Burgess Shale, his administrative burdens would not have allowed him time to study the Burgess fossils with anything like the requisite care and attention.

I have labored through the details of Walcott's interpretation and its sources because I know no finer illustration of the most important message taught by the history of science: the subtle and inevitable hold that theory exerts upon data and observation. Reality does not speak to us objectively, and no scientist can be free from constraints of psyche and society. The greatest impediment to scientific innovation is usually a conceptual lock, not a factual lack.

The transition from Walcott to Whittington is a premier example of this theme. The new view—as important an innovation as paleontology has ever contributed to our understanding of life and its history—was in no way closed to Walcott. Whittington and colleagues studied Walcott's specimens, using techniques and tools fully available in Walcott's time, in making their radical revision. They did not succeed as self-conscious revolutionaries, touting a new view in *a priori* assault. They began with Walcott's basic interpretation, but forged ahead on both sides of the great dialectic between theory and data—because they took the time to converse adequately with the Burgess fossils, and because they were willing to listen.

The transition from Walcott to Whittington marks a milestone that could hardly be exceeded in importance. The new view of the Burgess Shale is no more nor less than the triumph of history itself as a favored principle for reading the evolution of life.

THE BURGESS SHALE AND THE NATURE OF HISTORY

Our language is full of phrases that embody the worst and most restrictive stereotype about science. We exhort our frustrated friends to be "scientific"—meaning unemotional and analytic—in approaching a vexatious problem. We talk about the "scientific method," and instruct schoolchildren in this supposedly monolithic and maximally effective path to natural knowledge, as if a single formula could unlock all the multifarious secrets of empirical reality.

Beyond a platitudinous appeal to open-mindedness, the "scientific method" involves a set of concepts and procedures tailored to the image of a man in a white coat twirling dials in a laboratory—experiment, quantification, repetition, prediction, and restriction of complexity to a few variables that can be controlled and manipulated. These procedures are powerful, but they do not encompass all of nature's variety. How should scientists operate when they must try to explain the results of history, those inordinately complex events that can occur but once in detailed glory? Many large domains of nature—cosmology, geology, and evolution among them—must be studied with the tools of history. The appropriate methods focus on narrative, not experiment as usually conceived.

The stereotype of the "scientific method" has no place for irreducible history. Nature's laws are defined by their invariance in space and time. The techniques of controlled experiment, and reduction of natural complexity to a minimal set of general causes, presuppose that all times can be treated alike and adequately simulated in a laboratory. Cambrian quartz is like modern quartz—tetrahedra of silicon and oxygen bound together at all corners. Determine the properties of modern quartz under controlled conditions in a laboratory, and you can interpret the beach sands of the Cambrian Potsdam Sandstone.

But suppose you want to know why dinosaurs died, or why mollusks flourished while *Wiwaxia* perished? The laboratory is not irrelevant, and may yield important insights by analogy. (We might, for example, learn something interesting about the Cretaceous extinction by testing the physiological tolerances of modern organisms, or even of dinosaur "models," under environmental changes proposed in various theories for this great dying.) But the restricted techniques of the "scientific method" cannot get to the heart of this singular event involving creatures long dead on an earth with climates and continental positions markedly different from today's. The resolution of history must be rooted in the reconstruction of past events themselves—in their own terms—based on narrative evidence of their own unique phenomena. No law guaranteed the demise of *Wiwaxia*, but some complex set of events conspired to assure this result— and we may be able to recover the causes if, by good fortune, sufficient evidence lies recorded in our spotty geological record. (We did not, until ten years ago, for example, know that the Cretaceous extinction corresponded in time with the probable impact of one or several extraterrestrial bodies upon the earth—though the evidence, in chemical signatures, had always existed in rocks of the right age.)

Historical explanations are distinct from conventional experimental results in many ways. The issue of verification by repetition does not arise because we are trying to account for uniqueness of detail that cannot, both by laws of probability and time's arrow of irreversibility, occur together again. We do not attempt to interpret the complex events of narrative by reducing them to simple consequences of natural law; historical events do not, of course, violate any general principles of matter and motion, but their occurrence lies in a realm of contingent detail. (The law of gravity tells us how an apple falls, but not why that apple fell at that moment, and why Newton happened to be sitting there, ripe for inspiration.) And the issue of prediction, a central ingredient in the stereotype, does not enter into a historical narrative. We can explain an event after it occurs, but contingency precludes its repetition, even from an identical starting point. (Custer was doomed after a thousand events conspired to isolate his troops, but start again in 1850 and he might never see Montana, much less Sitting Bull and Crazy Horse.)

These differences place historical, or narrative, explanations in an unfavorable light when judged by restrictive stereotypes of the "scientific method." The sciences of historical complexity have therefore been demoted in status and generally occupy a position of low esteem among professionals. In fact, the status ordering of the sciences has become so familiar a theme that the ranking from adamantine physics at the pinnacle down to such squishy and subjective subjects as psychology and sociology at the bottom has become stereotypical in itself. These distinctions have entered our language and our metaphors—the "hard" versus the "soft" sciences, the "rigorously experimental" versus the "merely descriptive." Several years ago, Harvard University, in an uncharacteristic act of educational innovation, broke conceptual ground by organizing the sciences according to procedural style rather than conventional discipline within the core curriculum. We did not make the usual twofold division into physical versus biological, but recognized the two styles just discussed—the experimental-predictive and the historical. We designated each category by a letter rather than a name. Guess which division became Science A, and which Science B? My course on the history of earth and life is called Science B-16.

Perhaps the saddest aspect of this linear ranking lies in the acceptance of inferiority by bottom dwellers, and their persistent attempt to ape inappropriate methods that may work higher up on the ladder. When the order itself should be vigorously challenged, and plurality with equality asserted in pride, too many historical scientists act like the prison trusty who, ever mindful of his tenuous advantages, outdoes the warden himself in zeal for preserving the status quo of power and subordination.

Thus, historical scientists often import an oversimplified caricature of "hard" science, or simply bow to pronouncements of professions with higher status. Many geologists accepted Lord Kelvin's last and most restrictive dates for a young earth, though the data of fossils and strata spoke clearly for more time. (Kelvin's date bore the prestige of mathematical formulae and the weight of physics, though the discovery of radioactivity soon invalidated Kelvin's premise that heat now rising from the earth's interior records the cooling of our planet from an initially molten state not long past.) Even more geologists rejected continental drift, despite an impressive catalogue of data on previous connections among continents, because physicists had proclaimed the lateral motion of continents impossible. Charles Spearman misused the statistical technique of factor analysis to designate intelligence as a single, measurable, physical thing in the head, and then rejoiced for psychology because "this Cinderella among the sciences has made a bold bid for the level of triumphant physics itself" (quoted in Gould, 1981, p. 263).

But historical science is not worse, more restricted, or less capable of achieving firm conclusions because experiment, prediction, and subsumption under invariant laws of nature do not represent its usual working methods. The sciences of history use a different mode of explanation, rooted in the comparative and observational richness of our data. We cannot see a past event directly, but science is usually based on inference, not unvarnished observation (you don't see electrons, gravity, or black holes either).

In no other way but this false ordering by status among the sciences can I understand the curious phenomenon that led me to write this book in the first place—namely, that the Burgess revision has been so little noticed by the public in general and also by scientists in other disciplines. Yes, I understand that science writers don't consult the Philosophical Transactions of the Royal Society, London, *and that hundred-page anatomical monographs can seem rather daunting to those unschooled in the jargon. But we cannot charge Whittington and colleagues with hiding the good news. They have also published in the general journals that science writers do read—principally* Science *and* Nature. *They have written half a dozen prominent "review articles" for scientific colleagues. They have also composed a good deal for general audiences, including articles for* Scientific American *and* Natural History, *and a popular guide for Parks Canada. They know the implications of their work, and they have tried to get the message across; others have also aided (I have written four essays on the Burgess Shale for* Natural History). *Why has the story not taken hold, or been regarded as momentous?*

An interesting contrast, hinting at a solution, might be drawn between the Burgess revision and the Alvarez theory linking the Cretaceous extinction to extraterrestrial impact. I regard these two as the most important paleontological discoveries of the past twenty years. I think that they are equal in significance and that they tell the same basic story (as illustrations of the extreme chanciness and contingency of life's history: decimate the Burgess differently and we never evolve; send those comets into harmless orbits and dinosaurs still rule the earth, precluding the rise of large mammals, including humans). I hold that both are now well documented, the Burgess revision probably better than the Alvarez claim. Yet the asymmetry of public attention has been astonishing. Alvarez's impact theory has graced the cover of Time, *been featured in several television documentaries, and been a subject of comment and controversy wherever science achieves serious discussion. Few nonprofessionals have ever heard of the Burgess Shale—making this book necessary.*

I do understand that part of this difference in attention simply reflects our parochial fascination with the big and the fierce. Dinosaurs are destined for more attention than two-inch "worms." But I believe that the major ingredient—particularly in the decision of science writers to avoid the Burgess Shale—lies with the stereotype of the scientific method, and the false ordering of sciences by status. Luis Alvarez, who died as I was writing this book, was a Nobel laureate and one of the most brilliant physicists of our century; he was, in short, a prince of science at the highest conventional grade. The evidence for his theory lies in the usual stuff of the laboratory—precise measurements made with expensive machinery on minute quantities of iridium. The impact theory has everything for public acclaim—white coats, numbers, Nobel renown, and location at the top of the ladder of status. The Burgess redescriptions, on the other hand, struck many observers as one funny thing after another—just descriptions of some previously unappreciated, odd animals from early in life's history.

I loved Luie Alvarez for the excitement that he injected into my field. Our personal relationship was warm, for I was one of the few paleontologists who liked what he had to say from the outset (though not always, in retrospect, for good reasons). Yet, de mortuis nil nisi bonum notwithstanding, I must report that Luie could also be part of the problem. I do appreciate his frustration with so many paleontologists who, caught by traditions of gradualism and terrestrial causation, never paid proper attention to his evidence. Yet Luie often lashed out at the entire profession, and at historical science in general, claiming, for example, in an already infamous interview with the New York Times, "I don't like to say bad things about paleontologists, but they're really not very good scientists. They're more like stamp collectors."

I give Luie credit for saying out loud what many scientists of the stereotype think but dare not say, in the interests of harmony. The common epithet linking historical explanation with stamp collecting represents the classic arrogance of a field that does not understand the historian's attention to comparison among detailed particulars, all different. This taxonomic activity is not equivalent to licking hinges and placing bits of colored paper in preassigned places in a book. The historical scientist focuses on detailed particulars—one funny thing after another—because their coordination and comparison permits us, by consilience of induction, to explain the past with as much confidence (if the evidence is good) as Luie Alvarez could ever muster for his asteroid by chemical measurement.

We shall never be able to appreciate the full range and meaning of science until we shatter the stereotype of ordering by status and understand the different forms of historical explanation as activities equal in merit to anything done by physics or chemistry. When we achieve this new taxonomic arrangement of plurality among the sciences, then, and only then, will the importance of the Burgess Shale leap out. We shall then finally understand that the answer to such questions as "Why can humans reason?" lies as much (and as deeply) in the quirky pathways of contingent history as in the physiology of neurons.

The firm requirement for all science—whether stereotypical or historical—lies in secure testability, not direct observation. We must be able to determine whether our hypotheses are definitely wrong or probably correct (we leave assertions of certainty to preachers and politicians). History's richness drives us to different methods of testing, but testability is our criterion as well. We work with our strength of rich and diverse data recording the consequences of past events; we do not bewail our inability to see the past directly. We search for repeated pattern, shown by evidence so abundant and so diverse that no other coordinating interpretation could stand, even though any item, taken separately, would not provide conclusive proof.

The great nineteenth-century philosopher of science William Whewell devised the word consilience, meaning "jumping together," to designate the confidence gained when many independent sources "conspire" to indicate a particular historical pattern. He called the strategy of coordinating disparate results from multifarious sources consilience of induction.

I regard Charles Darwin as the greatest of all historical scientists. Not only did he develop convincing evidence for evolution as the coordinating principle of life's history, but he also chose as a conscious and central theme for all his writings—the treatises on worms, coral reefs, and orchids, as well as the great volumes on evolution—the development of a different but equally rigorous methodology for historical science (Gould, 1986). Darwin explored a variety of modes for historical explanation, each appropriate for differing densities of preserved information (Gould, 1986, pp. 60–64), but his central argument rested on Whewell's consilience. We know that evolution must underlie the order of life because no other explanation can coordinate the disparate data of embryology, biogeography, the fossil record, vestigial organs, taxonomic relationships, and so on. Darwin explicitly rejected the naive but widely held notion that a cause must be seen directly in order to qualify as a scientific explanation. He wrote about the proper testing of natural selection, invoking the idea of consilience for historical explanation:

> Now this hypothesis may be tested—and this seems to me the only fair and legitimate manner of considering the whole question—by trying whether it explains several large and independent classes of facts; such as the geological succession of organic beings, their distribution in past and present times, and their mutual affinities and homologies. If the principle of natural selection does explain these and other large bodies of facts, it ought to be received (1868, vol. 1, p. 657).

But historical scientists must then proceed beyond the simple demonstration that their explanations can be tested by equally rigorous procedures different from the stereotype of the "scientific method"; they must also convince other scientists that explanations of this historical type are both interesting and vitally informative. When we have established "just history" as the only complete and acceptable explanation for phenomena that everyone judges important—the evolution of the human intelligence, or of any self-conscious life on earth, for example—then we shall have won.

Historical explanations take the form of narrative: E, the phenomenon to be explained, arose because D came before, preceded by C, B, and A. If any of these earlier stages had not occurred, or had transpired in a different way, then E would not exist (or would be present in a substantially altered form, E', requiring a different explanation). Thus, E makes sense and can be explained rigorously as the outcome of A through D. But no law of nature enjoined E; any variant E' arising from an altered set of antecedents, would have been equally explicable, though massively different in form and effect.

I am not speaking of randomness (for E had to arise, as a consequence of A through D), but of the central principle of all history—*contingency*. A historical explanation does not rest on direct deductions from laws of nature, but on an unpredictable sequence of antecedent states, where any major change in any step of the sequence would have altered the final result. This final result is therefore dependent, or contingent, upon everything that came before—the unerasable and determining signature of history.

Many scientists and interested laypeople, caught by the stereotype of the "scientific method," find such contingent explanations less interesting or less "scientific," even when their appropriateness and essential correctness must be acknowledged. The South lost the Civil War with a kind of relentless inevitability once hundreds of particular events happened as they did—Pickett's charge failed, Lincoln won the election of 1864, *etc., etc., etc.* But wind the tape of American history back to the Louisiana Purchase, the Dred Scott decision, or even only to Fort Sumter, let it run again with just a few small and judicious changes (plus their cascade of consequences), and a different outcome, including the opposite resolution, might have occurred with equal relentlessness past a certain point. (I used to believe that Northern superiority in population and industry had virtually guaranteed the result from the start. But I have been persuaded by recent scholarship that wars for recognition rather than conquest can be won by purposeful minorities. The South was not trying to overrun the North, but merely to secure its own declared borders and win acknowledgment as an independent state. Majorities, even in the midst of occupation, can be rendered sufficiently war-weary and prone to withdraw by insurgencies, particularly in guerilla form, that will not relent.)

Suppose, then, that we have a set of historical explanations, as well documented as anything in conventional science. These results do not arise as deducible consequences from any law of nature; they are not even predictable from any general or abstract property of the larger system (as superiority in population or industry). How can we deny such explanations a role every bit as interesting and important as a more conventional scientific conclusion? I hold that we must grant equal status for three basic reasons.

1. *A question of reliability.* The documentation of evidence, and probability of truth by disproof of alternatives, may be every bit as conclusive as for any explanation in traditional science.

2. *A matter of importance.* The equal impact of historically contingent explanations can scarcely be denied. The Civil War is the focus and turning point of American history. Such central matters as race, regionalism, and economic power owe their present shape to this great event that need not have occurred. If the current taxonomic order and relative diversity of life are more a consequence of "just history" than a potential deduction from general principles of evolution, then contingency sets the basic pattern of nature.

3. *A psychological point.* I have been too apologetic so far. I have even slipped into the rhetoric of inferiority—by starting from the premise that historical explanations may be less interesting and then pugnaciously fighting for equality. No such apologies need be made. Historical explanations are endlessly fascinating in themselves, in many ways more intriguing to the human psyche than the inexorable consequences of nature's laws. We are especially moved by events that did not have to be, but that occurred for identifiable reasons subject to endless mulling and stewing. By contrast, both ends of the usual dichotomy—the inevitable and the truly random—usually make less impact on our emotions because they cannot be controlled by history's agents and objects, and are therefore either channeled or buffeted, without much hope for pushing back. But, with contingency, we are drawn in; we become involved; we share the pain of triumph or tragedy. When we realize that the actual outcome did not have to be, that any alteration in any step along the way would have unleashed a cascade down a different channel, we grasp the causal power of individual events. We can argue, lament, or exult over each detail—because each holds the power of transformation. Contingency is the affirmation of control by immediate events over destiny, the kingdom lost for want of a horseshoe nail. The Civil War is an especially poignant tragedy because a

replay of the tape might have saved a half million lives for a thousand different reasons—and we would not find a statue of a soldier, with names of the dead engraved on the pedestal below, on every village green and before every county courthouse in old America. Our own evolution is a joy and a wonder because such a curious chain of events would probably never happen again, but having occurred, makes eminent sense. Contingency is a license to participate in history, and our psyche responds.

The theme of contingency, so poorly understood and explored by science, has long been a mainstay of literature. We note here a situation that might help to breach the false boundaries between art and nature, and even allow literature to enlighten science. Contingency is Tolstoy's cardinal theme in all his great novels. Contingency is the source of tension and intrigue in many fine works of suspense, most notably in a recent masterpiece by Ruth Rendell (writing as Barbara Vine), *A Fatal Inversion* (1987)—a chilling book describing a tragedy that engulfs the lives and futures of a small community through an escalating series of tiny events, each peculiar and improbable (but perfectly plausible) in itself, and each entraining a suite of even stranger consequences. *A Fatal Inversion* is so artfully and intricately plotted by this device that I must view Rendell's finest work as a conscious text on the nature of history.

Two popular novels of the past five years have selected Darwinian theory as their major theme. I am especially intrigued and pleased that both accept and explore contingency as the theory's major consequence for our lives. In this correct decision, Stephen King and Kurt Vonnegut surpass many scientists in their understanding of evolution's deeper meanings.

King's *The Tommyknockers* (1987) fractures a tradition in science fiction by treating extraterrestrial "higher intelligences" not as superior in general, wiser, or more powerful, but merely as quirky hangers-on in the great Darwinian game of adaptation by differential reproductive success in certain environments. (King refers to this persistence as "dumb evolution"; I just call it Darwinism.)* Such equivocal success by endless and immediate adjustment breeds contingency, which then becomes the controlling theme of *The Tommyknockers*—as the aliens fail in their plans for earth, thanks largely to evasive action by one usually ineffective, cynical, and dipsomaniacal English professor. King muses on the nature of controlling events in contingent sequences, and on their level of perceived importance at various scales:

*Our agreement on the theme, if not the terminology, provides hope that even the most implacable differences in style and morality may find a common meeting ground on this most important of intellectual turfs—for Steve is the most fanatical Red Sox booster in New England, while my heart remains with the Yankees.

I would not be the one to tell you there are no planets anywhere in the universe that are not large dead cinders floating in space because a war over who was or was not hogging too many dryers in the local Laundromat escalated into Doomsville. No one ever really knows where things will end—or if they will. . . . Of course we may blow up our world someday with no outside help at all, for reasons which look every bit as trivial from a standpoint of light-years; from where we rotate far out on one spoke of the Milky Way in the Lesser Magellanic Cloud, whether or not the Russians invade the Iranian oilfields or whether NATO decides to install American-made Cruise missiles in West Germany may seem every bit as important as whose turn it is to pick up the tab for five coffees and a like number of Danish.

Kurt Vonnegut's *Galápagos* (1985) is an even more conscious and direct commentary on the meaning of evolution from a writer's standpoint. I feel especially gratified that a cruise to the Galápagos, a major source of Vonnegut's decision to write the book, should have suggested contingency as the cardinal theme taught by Darwin's geographic shrine. In Vonnegut's novel, the pathways of history may be broadly constrained by such general principles as natural selection, but contingency has so much maneuvering room within these boundaries that any particular outcome owes more to a quirky series of antecedent events than to channels set by nature's laws. *Galápagos*, in fact, is a novel about the nature of history in Darwin's world. I would (and do) assign it to students in science courses as a guide to understanding the meaning of contingency.

In *Galápagos*, the holocaust of depopulation arrives by the relatively mild route of a bacterium that destroys human egg cells. This scourge first gains a toehold by striking women at the annual international book fair in Frankfurt, but quickly spreads throughout the world, sterilizing all but an isolated remnant of *Homo sapiens*. Human survival becomes concentrated in a tiny and motley group carried by boat beyond the reach of the bacterium to the isolated Galápagos—the last of the Kanka-bono Indians plus a tourist and adventurer or two. Their survival and curious propagation proceeds through a wacky series of contingencies, yet all future human history now resides with this tiny remnant:

> In a matter of less than a century the blood of every human being on earth would be predominantly Kanka-bono, with a little von Kleist and Hiroguchi thrown in. And this astonishing turn of events would be made to happen, in large part, by one of the only two absolute nobodies on the original passenger list for "the Nature Cruise of the Century." That was Mary Hepburn. The other nobody was her husband, who himself played a crucial role in shaping human destiny by booking, when facing his own extinction, that one cheap little cabin below the waterline.

Contingency has also been an important theme in films, both recent and classic. In *Back to the Future* (1985) Marty McFly (Michael J. Fox), a teen-ager transported back in time to the high school attended by his parents, must struggle to reconstitute the past as it actually happened, after his accidental intrusion threatens to alter the initial run of the tape (when his mother, in an interesting variation on Oedipus, develops a crush on him). The events that McFly must rectify seem to be tiny occurrences of absolutely no moment, but he knows that nothing could be more important, since failure will result in that ultimate of consequences, his own erasure, because his parents will never meet.

The greatest expression of contingency—my nomination as the holotype* of the genre—comes near the end of Frank Capra's masterpiece, *It's a Wonderful Life* (1946). George Bailey (Jimmy Stewart) has led a life of self-abnegation because his basic decency made him defer personal dreams to offer support for family and town. His precarious building and loan association has been driven to bankruptcy and charged with fraud through the scheming of the town skinflint and robber baron, Mr. Potter (Lionel Barrymore). George, in despair, decides to drown himself, but Clarence Odbody, his guardian angel, intervenes by throwing himself into the water first, knowing that George's decency will demand another's rescue in preference to immediate suicide. Clarence then tries to cheer George up by the direct route: "You just don't know all that you've done"; but George replies: "If it hadn't been for me, everybody'd be a lot better off. . . . I suppose it would have been better if I'd never been born at all."

Clarence, in a flash of inspiration, grants George his wish and shows him an alternative version of life in his town of Bedford Falls, replayed in his complete absence. This magnificent ten-minute scene is both a highlight of cinematic history and the finest illustration that I have ever encountered for the basic principle of contingency—a replay of the tape yielding an entirely different but equally sensible outcome; small and apparently insignificant changes, George's absence among others, lead to cascades of accumulating difference.

Everything in the replay without George makes perfect sense in terms of personalities and economic forces, but this alternative world is bleak and cynical, even cruel, while George, by his own apparently insignificant life,

---

*"Holotype" is taxonomic jargon for the specimen designated to bear the name of a species. Holotypes are chosen because concepts of the species may change later and biologists must have a criterion for assigning the original name. (If, for example, later taxonomists decide that two species were mistakenly mixed together in the first description, the original name will go to the group including the holotype specimen.)

had imbued his surroundings with kindness and attendant success for his beneficiaries. Bedford Falls, his idyllic piece of small-town America, is now filled with bars, pool halls, and gambling joints; it has been renamed Pottersville, because the Bailey Building and Loan failed in George's absence and his unscrupulous rival took over the property and changed the town's name. A graveyard now occupies the community of small homes that George had financed at low interest and with endless forgiveness of debts. George's uncle, in despair at bankruptcy, is in an insane asylum; his mother, hard and cold, runs a poor boarding house; his wife is an aging spinster working in the town library; a hundred men lay dead on a sunken transport, because his brother drowned without George to rescue him, and never grew up to save the ship and win the Medal of Honor.

The wily angel, clinching his case, then pronounces the doctrine of contingency: "Strange, isn't it? Each man's life touches so many other lives, and when he isn't around he leaves an awful hole, doesn't he? . . . You see, George, you really had a wonderful life."

Contingency is both the watchword and lesson of the new interpretation of the Burgess Shale. The fascination and transforming power of the Burgess message—a fantastic explosion of early disparity followed by decimation, perhaps largely by lottery—lies in its affirmation of history as the chief determinant of life's directions.

Walcott's earlier and diametrically opposite view located the pattern of life's history firmly in the other and more conventional style of scientific explanation—direct predictability and subsumption under invariant laws of nature. Moreover, Walcott's view of invariant law would now be dismissed as more an expression of cultural tradition and personal preference than an accurate expression of nature's patterns. For as we have seen, Walcott read life's history as the fulfillment of a divine purpose guaranteed to yield human consciousness after a long history of gradual and stately progress. The Burgess organisms had to be primitive versions of later improvements, and life had to move forward from this restricted and simple beginning.

The new view, on the other hand, is rooted in contingency. With so many Burgess possibilities of apparently equivalent anatomical promise— over twenty arthropod designs later decimated to four survivors, perhaps fifteen or more unique anatomies available for recruitment as major branches, or phyla, of life's tree—our modern pattern of anatomical disparity is thrown into the lap of contingency. The modern order was not guaranteed by basic laws (natural selection, mechanical superiority in anatomical design), or even by lower-level generalities of ecology or evolutionary theory. The modern order is largely a product of contingency. Like

Bedford Falls with George Bailey, life had a sensible and resolvable history, generally pleasing to us since we did manage to arise, just a geological minute ago. But, like Pottersville without George Bailey, any replay, altered by an apparently insignificant jot or tittle at the outset, would have yielded an equally sensible and resolvable outcome of entirely different form, but most displeasing to our vanity in the absence of self-conscious life. (Though, needless to say, our nonexistent vanity would scarcely be an issue in any such alternative world.) By providing a maximum set of anatomically proficient possibilities right at the outset, the Burgess Shale becomes our centerpiece for the controlling power of contingency in setting the pattern of life's history and current composition.

Finally, if you will accept my argument that contingency is not only resolvable and important, but also fascinating in a special sort of way, then the Burgess not only reverses our general ideas about the source of pattern—it also fills us with a new kind of amazement (also a *frisson* for the improbability of the event) at the fact that humans ever evolved at all. We came *this close* (put your thumb about a millimeter away from your index finger), thousands and thousands of times, to erasure by the veering of history down another sensible channel. Replay the tape a million times from a Burgess beginning, and I doubt that anything like *Homo sapiens* would ever evolve again. It is, indeed, a wonderful life.

A final point about predictability versus contingency: Am I really arguing that nothing about life's history could be predicted, or might follow directly from general laws of nature? Of course not; the question that we face is one of scale, or level of focus. Life exhibits a structure obedient to physical principles. We do not live amidst a chaos of historical circumstance unaffected by anything accessible to the "scientific method" as traditionally conceived. I suspect that the origin of life on earth was virtually inevitable, given the chemical composition of early oceans and atmospheres, and the physical principles of self-organizing systems. Much about the basic form of multicellular organisms must be constrained by rules of construction and good design. The laws of surfaces and volumes, first recognized by Galileo, require that large organisms evolve different shapes from smaller relatives in order to maintain the same relative surface area. Similarly, bilateral symmetry can be expected in mobile organisms built by cellular division. (The Burgess weird wonders are bilaterally symmetrical.)

But these phenomena, rich and extensive though they are, lie too far from the details that interest us about life's history. Invariant laws of nature impact the general forms and functions of organisms; they set the channels in which organic design must evolve. But the channels are so broad relative to the details that fascinate us! The physical channels do not specify arthropods, annelids, mollusks, and vertebrates, but, at most, bilaterally symmetrical organisms based on repeated parts. The boundaries of the channels retreat even further into the distance when we ask the essential questions about our own origin: Why did mammals evolve among vertebrates? Why did primates take to the trees? Why did the tiny twig that produced *Homo sapiens* arise and survive in Africa? When we set our focus upon the level of detail that regulates most common questions about the history of life, contingency dominates and the predictability of general form recedes to an irrelevant background.

Charles Darwin recognized this central distinction between *laws in the background* and *contingency in the details* in a celebrated exchange of letters with the devout Christian evolutionist Asa Gray. Gray, the Harvard botanist, was inclined to support not only Darwin's demonstration of evolution but also his principle of natural selection as its mechanism. But Gray was worried about the implications for Christian faith and the meaning of life. He particularly fretted that Darwin's view left no room for rule by law, and portrayed nature as shaped entirely by blind chance.

Darwin, in his profound reply, acknowledged the existence of general laws that regulate life in a broad sense. These laws, he argued, addressing Gray's chief concern, might even (for all we know) reflect some higher purpose in the universe. But the natural world is full of details, and these form the primary subject matter of biology. Many of these details are "cruel" when measured, inappropriately, by human moral standards. He wrote to Gray: "I cannot persuade myself that a beneficent and omnipotent God would have designedly created the Ichneumonidae with the express intention of their feeding within the living bodies of Caterpillars, or that a cat should play with mice." How, then, could the nonmorality of details be reconciled with a universe whose general laws might reflect some higher purpose? Darwin replied that the details lay in a realm of contingency undirected by laws that set the channels. The universe, Darwin replied to Gray, runs by law, "with the details, whether good or bad, left to the working out of what we may call chance."

And so, ultimately, the question of questions boils down to the placement of the boundary between predictability under invariant law and the multifarious possibilities of historical contingency. Traditionalists like Walcott would place the boundary so low that all major patterns of life's history fall above the line into the realm of predictability (and, for him, direct manifestation of divine intentions). But I envision a boundary sitting so high that almost every interesting event of life's history falls into the realm of contingency. I regard the new interpretation of the Burgess Shale as nature's finest argument for placing the boundary this high.

This means—and we must face the implication squarely—that the origin of *Homo sapiens*, as a tiny twig on an improbable branch of a contingent limb on a fortunate tree, lies well below the boundary. In Darwin's scheme, we are a detail, not a purpose or embodiment of the whole—"with the details, whether good or bad, left to the working out of what we may call chance." Whether the evolutionary origin of self-conscious intelligence in any form lies above or below the boundary, I simply do not know. All we can say is that our planet has never come close a second time.

For anyone who feels cosmically discouraged at the prospect of being a detail in the realm of contingency, I cite for solace a wonderful poem by Robert Frost, dedicated explicitly to this concern: *Design*. Frost, on a morning walk, finds an odd conjunction of three white objects with different geometries. This peculiar but fitting combination, he argues, must record some form of intent; it cannot be accidental. But if intent be truly manifest, then what can we make of our universe—for the scene is evil by any standard of human morality. We must take heart in Darwin's proper solution. We are observing a contingent detail, and may yet hope for purpose, or at least neutrality, from the universe in general.

> I found a dimpled spider, fat and white,
> On a white heal-all, holding up a moth
> Like a white piece of rigid satin cloth—
> Assorted characters of death and blight
> Mixed ready to begin the morning right,
> Like the ingredients of a witches' broth—
> A snow-drop spider, a flower like a froth,
> And dead wings carried like a paper kite.
>
> What had that flower to do with being white,
> The wayside blue and innocent heal-all?
> What brought the kindred spider to that height,
> Then steered the white moth thither in the night?
> What but design of darkness to appall?—
> If design govern in a thing so small.

Homo sapiens, I fear, is a "thing so small" in a vast universe, a wildly improbable evolutionary event well within the realm of contingency. Make of such a conclusion what you will. Some find the prospect depressing; I have always regarded it as exhilarating, and a source of both freedom and consequent moral responsibility.

CHAPTER V

Possible Worlds: The Power of "Just History"

A STORY OF ALTERNATIVES

In the last chapter I gave the general, abstract brief for contingency. But the case for "just history" cannot rest on mere plausibility or force of argument. I must be able to convince you—by actual example—that honorable, reasonable, and fascinatingly different alternatives could have produced a substantially divergent history of life not graced by human intelligence.

The problem, of course, with describing alternatives is that they didn't happen—and we cannot know the details of their plausible occurrence. I feel certain, for example, that no Burgess paleontologist could have surveyed the twenty-five possibilities of arthropod design, rejected the most common (and anatomically sleek) *Marrella*, put aside the beautifully complex *Leanchoilia* or the sturdy, workaday *Sidneyia*, and admitted the ecologically specialized *Aysheaia* and the rare *Sanctacaris* to the company of the elect. But even if we could envision a modern arthropod world built by descendants of *Marrella*, *Leanchoilia*, and *Sidneyia*, how could we specify the forms that their descendants would take? After all, we cannot even make predictions when we know the line of descent: we cannot see the mayfly in *Aysheaia*, or the black widow spider in *Sanctacaris*. How can we specify the world that different decimations would have produced?

I believe that the best response to this dilemma is to adopt a more modest approach. Instead of seeking an illustration based on unknowable descendants of groups that did not in fact survive, let us consider a plausible alternative world different only in the diversity of two groups that graced the Burgess and survive today—for here we need conjecture only about the reasons for relative abundance. Take two groups of modern oceans—one bursting with diversity, the other nearly gone. Would we have known, at the Burgess beginning of both, which was destined for domination and which for peripheral status in the nooks and crannies of an unforgiving world? Can we make a plausible case for a replay with opposite outcome? (Again, as for so much of this book, I owe this example to the suggestion and previous probing of Simon Conway Morris.)

Consider the current distribution of two phyla sharing the most common invertebrate body plan—the flexible, elongate, bilateral symmetry of "worms." Polychaetes, the major marine component of the phylum Annelida (including earthworms on land), represent one of life's great success stories. The best modern epitome, Sybil P. Parker's *McGraw-Hill Synopsis and Classification of Living Organisms* (1982), devotes forty pages to a breathless summary of their eighty-seven families, one thousand genera, and some eight thousand species. Polychaetes range in size from less than one millimeter to more than three meters; they live nearly everywhere, most on the sea floor, but some in brackish or fresh water, and a few in moist earth. Their life styles also span the range of the thinkable: most are free-living and carnivorous or scavenging, but others dwell commensally with sponges, mollusks, or echinoderms, and some are parasites.

By contrast, consider the priapulids, burrowing worms with bodies divided roughly into three parts—a rear end with one or two appendages, a middle trunk, and a retractable front end, or proboscis. Both the form of the proboscis and its power of erection from the trunk inevitably reminded early male zoologists of something else to which they were, no doubt, firmly and fondly attached—hence the burden of nomenclature for these creatures as *Priapulus*, or the "little penis."

The armature of the priapulid proboscis might give some cause for alarm in unwarranted analogy. In most species the lower portion sports twenty-five rows of little teeth, or scalids, surmounted by a collar, or buccal ring. The upper end contains several inscribed pentagons of teeth surrounding the mouth. Most priapulids are active carnivores, capturing and swallowing their prey whole, although one species may feed on detritus.

But when we turn to Parker's compendium of living organisms, we find but three pages devoted to priapulids, with a leisurely description of each family. Priapulids just don't contribute much to an account of organic diversity; zoologists have found only about fifteen species. For some reason, priapulids do not rank among the success stories of modern biology.

An examination of priapulid distribution provides a clue to their relative failure. All priapulids live in unusual, harsh, or marginal environments—as if they cannot compete in the shallow, open environments frequented by most "standard" marine organisms, and can hang on only where ordinary creatures don't bother. Two priapulid families include worms grown so small that they live among sand grains in the rich and fascinating (but decidedly "unstandard") world of the so-called interstitial fauna. Most priapulids belong to the family Priapulidae, larger worms (up to twenty centimeters) of the sea bottom. But these priapulids do not inhabit the richest environments of the shallow-water tropics. They live in the coldest realms—at great depths in tropical regions, and in shallow waters in the frigid climates of high latitudes. They can also tolerate a variety of unusual conditions—low oxygen levels, hydrogen sulfide, low or sharply fluctuating salinity, and unproductive surroundings that impose long periods of starvation. It does not strain the boundaries of reasonable inference to argue that priapulids have managed to keep a toehold in a tough world by opting for difficult places devoid of sharp competition.

We might assume that these striking differences between modern polychaetes and priapulids indicate something so intrinsic about the relative mettle of these two groups that their geological history should be an uninterrupted tale of polychaete prosperity and priapulid struggle. If so, we are in for yet another surprise from the redoubtable Burgess fauna. This first recorded beginning of modern soft-bodied life contains six genera of polychaetes and six or seven genera of priapulids. (See Conway Morris's monographs on priapulids, 1977d and polychaetes, 1979.)

Furthermore, the Burgess priapulids are numerically a major component of the fauna and, along with anomalocarids and a few arthropods, the earth's first important soft-bodied carnivores. *Ottoia prolifica* (figure 5.1), most common of the Burgess priapulids, swallowed its prey whole. Hyolithids (conical shelled creatures of uncertain affinity) were favored as food. Thirty-one specimens have been found in the guts of *Ottoia*, most swallowed in the same orientation (and, therefore, almost certainly hunted and consumed in a definite style). One *Ottoia* had six hyolithids in its gut. Another specimen had eaten some of its own—the earliest example of cannibalism in the fossil record.

By contrast, polychaetes (figure 5.2), though equal to priapulids in taxonomic diversity, are much rarer numerically. Conway Morris remarks: "In comparison with the situation in many modern marine environments, the Burgess Shale polychaetes had a relatively minor role."

Obviously, something dramatic (and disastrous) has happened to pria-

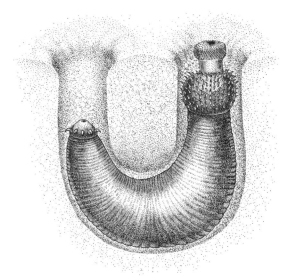

5.1. The Burgess priapulid *Ottoia* in its burrow, with its proboscis half extended. Drawn by Marianne Collins.

pulids since the Burgess. Once, they had no rivals for abundance among soft-bodied forms, exceeding even the proud polychaetes of current majesty. Now, they are few and forgotten, denizens of the ocean's spatial and environmental peripheries. The entire modern world contains scarcely more genera of priapulids than the single Burgess fauna from one quarry in British Columbia—while Burgess priapulids occupied center stage, not the tawdry provinces. What happened?

We do not know. It is tempting to argue that polychaetes had some biological leverage from the start and were destined for domination, however modest their beginning. But we have no idea what such an advantage might be. Conway Morris makes the intriguing observation that Burgess polychaetes had no jaws and that these organs of successful polychaete predators did not evolve until the subsequent Ordovician period. Perhaps the origin of jaws gave polychaetes their edge over the previously more abundant priapulids?

This supposition is plausible and may be correct, but we do not know; and a correlation (jaws with the beginning of dominance) need not imply a cause. In any case, our hypothetical Burgess geologist would not have known that the modest polychaetes would evolve jaws fifty million years hence.

5.2. The Burgess polychaete *Canadia.* Drawn by Marianne Collins.

The distribution and scarcity of modern priapulids, relative to Burgess abundance, does indicate a basic failure, but who can reconstruct the whys or wherefores? And who can say that a replay of life's tape would not yield a modern world dominated by priapulids, with a few struggling jawless polychaetes at a tenuous periphery? What did happen makes sense; our world is not capricious. But many other plausible scenarios would have satisfied any modern votary of progress and good sense, and priapulid dominance lies firmly among the might-have-beens.

Are these Burgess fancies common to life's history throughout or an oddity of uncertain beginnings, superseded by later inexorability? Consider one more might-have-been: When dinosaurs perished in the Cretaceous debacle, they left a vacuum in the world of large-bodied carnivores. Did the current reign of cats and dogs emerge by predictable necessity or contingent fortune? Would an Eocene paleontologist, surveying the vertebrate world fifty million years ago, have singled out for success the ancestors of Leo, king of beasts?

I doubt it. The Eocene world sported many lineages of mammalian carnivores, only one ancestral to modern forms and not especially distinguished at the time. But the Eocene featured a special moment in the history of carnivores, a pivot between two possibilities—one realized, the other forgotten. Mammals did not hold all the chips. In 1917, the American paleontologists W. D. Matthew and W. Granger described a "magnificent and quite unexpected" skeleton of a giant predacious bird from the Eocene of Wyoming, *Diatryma gigantea:*

Diatryma was a gigantic bird, ground living and with vestigial wings. In bulk of body and limbs it equalled all but the largest of moas and surpassed any living bird. . . . The height of the reconstructed skeleton is nearly 7 feet. The neck and head were totally unlike any living bird, the neck short and very massive, the head of enormous size with a huge compressed beak (1917).

The gigantic head and short, powerful neck identify *Diatryma* as a fierce carnivore, in sharp contrast with the small head and long, slender neck of the more peaceful ratites (ostriches, rheas, and their relatives). Like *Tyrannosaurus*, with its diminutive forelimbs but massive head and powerful hind limbs, *Diatryma* must have kicked, clawed, and bitten its prey into submission.

Diatrymids, distant relatives perhaps of cranes but no kin to ostriches and their ilk, ranged over Europe and North America for several million years. The plum of dominant carnivory could have fallen to the birds, but mammals finally prevailed, and we do not know why. We can invent stories about two legs, bird brains, and no teeth as necessarily inferior to all fours and sharp canines, but we know in our heart of hearts that if birds had won, we could tell just as good a tale about their inevitable success. A. S. Romer, leading vertebrate paleontologist of the generation just past, wrote in his textbook, the bible of the profession:

> The presence of this great bird at a time when mammals were, for the most part, of very small size (the contemporary horse was the size of a fox terrier) suggests some interesting possibilities—which never materialized. The great reptiles had died off, and the surface of the earth was open for conquest. As possible successors there were the mammals and the birds. The former succeeded in the conquest, but the appearance of such a form as *Diatryma* shows that the birds were, at the beginning, rivals of the mammals (1966, p. 171).

In all these speculations about replaying life's tape, we lament our lack of any controlled experiment. We cannot instigate the actual replay, and our planet provided only one run-through. But the crucial Eocene pivot between birds and mammals provides more and different evidence. For once, our recalcitrant and complex planet actually performed a proper experiment for us. This particular tape did have a replay, in South America—and this time the birds won, or at least held the mammals to a respectable draw!

South America was an island continent, a kind of super-Australia, until the Isthmus of Panama arose just a few million years ago. Most animals usually considered as distinctively South American—jaguars, llamas, and tapirs, for example—are North American migrants of postisthmian arrival.

The great native fauna of South America is largely gone (or surviving as a poor, if fascinating, remnant of armadillos, sloths, and the "Virginia" opossum, among others). No placental carnivores inhabited this giant ark. Most popular books tell us that the native South American carnivores were all marsupials, the so-called borhyaenids. They often neglect to say that another prominent group—the phororhacids, giant ground birds—fared just as well, if not better. Phororhacids also sported large heads and short, stout necks, but were not closely related to *Diatryma.* In South America, birds had a second and separate try as dominant carnivores, and this time they won, as suggested in Charles R. Knight's famous reconstruction of a phororhacid standing in triumph over a mammalian victim (figure 5.3).

In our smug, placental-centered parochialism, we may say that birds could triumph in South America only because marsupials are inferior to placentals and did not offer the kind of challenge that conquered predacious ground birds in Europe and North America. But can we be so sure? Borhyaenids could also be large and fierce, ranging to bear size and including such formidable creatures as *Thylacosmilus,* the marsupial sabertooth.

5.3.   A phororhacid bird of South America stands in triumph over its mammalian prey in this depiction by Charles R. Knight.

We might also sneer and point out that, in any case, phororhacids quickly snuffed it (along with borhyaenids) as soon as superior placentals flooded over the rising isthmus. But this common saga of progress will not wash either. G. G. Simpson, our greatest expert on the evolution of South American mammals, wrote in one of his last books:

> It has sometimes been said that these and other flightless South American birds . . . survived because there were long no placental carnivores on that continent. That speculation is far from convincing. . . . Most of the phororhacids became extinct before, only a straggler or two after, placental carnivores reached South America. Many of the borhyaenids that lived among these birds for many millions of years were highly predacious. . . . The phororhacids . . . were more likely to kill than to be killed by mammals (Simpson, 1980, pp. 147–50).

We must conclude, I think, that South America does represent a legitimate replay—round two for the birds.

GENERAL PATTERNS THAT ILLUSTRATE CONTINGENCY

This story of worms and birds—the first part graced with the sweep of history from Burgess times to now, the second with the virtues of repetition by natural experiment—moves contingency from a general statement about history into the realm of tangible things. A single story can establish plausibility by example, but it cannot make a complete case. The argument of this book needs two final supports: first, a statement about general properties of life's history that reinforce the claims of contingency; and second, a chronology of examples illustrating the power of contingency not for selected and specific cases alone, but for the most general pathways and probabilities of life on our planet. This section and the next present these final supports for my argument; an epilogue on an arresting fact then completes the book.

If geological time had operated exactly as Darwin envisioned, contingency would still reign, with perhaps a bit more of life's general pattern thrown into the realm of predictability under broad principles. Remember that Darwin viewed the history of life through his controlling metaphors of competition and the wedge (see page 229): the world is full of species, wedges crowded together on a log, and new forms can enter ecological communities only by displacing others (popping the wedges out). Displacement proceeds by competition under natural selection, and the better-adapted species win. Darwin felt that this process, operating in the micro-moment of the here and now, could be extrapolated into the countless millennia of geological time to yield the overall pattern of life's history. For example, in chapter 10 of the *Origin of Species,* Darwin labored mightily (if incorrectly, in retrospect) to show that extinctions are not rapid and simultaneous across large differences of form and environment, but that each major group peters out slowly, its decline linked with the rise of a superior competitor.* But by "better adapted," Darwin only meant "more suited to changing local environments," not superior in any general anatomical sense. The pathways to local adaptation are as likely to restrict as to enhance the prospects for long-term success (simplification in parasites, overelaboration in peacocks). Moreover, nothing else is as quirky and unpredictable—both in our metaphors and on our planet—as trends in climate and geography. Continents fragment and disperse; oceanic circulation changes; rivers alter their course; mountains rise; estuaries dry up. If life works more by tracking environment than by climbing up a ladder of progress, then contingency should reign.

I assert the powerful role of contingency in Darwin's system not as a logical corollary of his theory, but as an explicit theme central to his own life and work. Darwin invoked contingency in a fascinating way as his primary support for the fact of evolution itself. He embedded his defense in a paradox: One might think that the best evidence for evolution would reside in those exquisite examples of optimal adaptation presumably wrought by natural selection—the aerodynamic perfection of a feather or the flawless mimicry of insects that look like leaves or sticks. Such phenomena provide our standard textbook examples for the power of evolutionary modification—the mills of natural selection may operate slowly, but they grind exceedingly fine. Yet Darwin recognized that perfection cannot provide evidence for evolution because optimality covers the tracks of history.

If feathers are perfect, they may as well have been designed from scratch by an omnipotent God as from previous anatomy by a natural process. Darwin recognized that the primary evidence for evolution must be sought in quirks, oddities, and imperfections that lay bare the pathways of history. Whales, with their vestigial pelvic bones, must have descended from terrestrial ancestors with functional legs. Pandas, to eat bamboo, must build

---

*Mass extinctions do not negate the principle of natural selection, for environments can change too fast and too profoundly for organic response; but coordinated dyings do run counter to Darwin's preference for seeing the large in the small, and for viewing organic competition, group by separate group, as the primary source of life's overall pattern.

an imperfect "thumb" from a nubbin of a wrist bone, because carnivorous ancestors lost the requisite mobility of their first digit. Many animals of the Galápagos differ only slightly from neighbors in Ecuador, though the climate of these relatively cool volcanic islands diverges profoundly from conditions on the adjacent South American mainland. If whales retained no trace of their terrestrial heritage, if pandas bore perfect thumbs, if life on the Galápagos neatly matched the curious local environments—then history would not inhere in the productions of nature. But contingencies of "just history" do shape our world, and evolution lies exposed in the panoply of structures that have no other explanation than the shadow of their past.

Thus, contingency rules even in Darwin's world of extrapolation from organic competition within local communities chock-full of species. However, an exciting intellectual movement of the last quarter century has led us to recognize that nature is not so smoothly and continuously ordered; the large does not emerge from the small simply by adding more time. Several large-scale patterns—based on the nature of macroevolution and the history of environments—impose their own signatures on nature's pathways, and also disrupt, reset, and redirect whatever may be accumulating through time by the ticking of processes in the immediate here and now. Most of these patterns strongly reinforce the theme of contingency (see Gould, 1985a). Let us consider just two.

THE BURGESS PATTERN OF MAXIMAL INITIAL PROLIFERATION

The major argument of this book holds that contingency is immeasurably enhanced by the primary insight won from the Burgess Shale—that current patterns were not slowly evolved by continuous proliferation and advance, but set by a pronounced decimation (after a rapid initial diversification of anatomical designs), probably accomplished with a strong, perhaps controlling, component of lottery.

But we must know if the Burgess represents an odd incident or a general theme in life's history—for if most evolutionary bushes look like Christmas trees, with maximal breadth at their bottoms, then contingency wins its greatest possible boost as a predominant force in the history of organic disparity. My feeling about the importance of this question has led me to devote much of my technical research during the past fifteen years to the prevalence of "bottom-heaviness" in evolutionary trees (Raup *et al.*, 1973; Raup and Gould, 1974; Gould *et al.*, 1977; Gould, Gilinsky, and German, 1987).

Paleontologists have long recognized the Burgess pattern of maximal early disparity in conventional groups of fossils with hard parts. The echinoderms provide our premier example. All modern representatives of this exclusively marine phylum fall into five major groups—the starfishes (Asteroidea), the brittle stars (Ophiuroidea), the sea urchins and sand dollars (Echinoidea), the sea lilies (Crinoidea), and the sea cucumbers (Holothuroidea). All share the basic pattern of fivefold radial symmetry. Yet Lower Paleozoic rocks, at the inception of the phylum, house some twenty to thirty basic groups of echinoderms, including some anatomies far outside the modern boundaries. The edrioasteroids built their globular skeletons in three-part symmetry. The bilateral symmetry of some "carpoids" is so pronounced that a few paleontologists view them as possible ancestors of fishes, and therefore of us as well (Jefferies, 1986). The bizarre helicoplacoids grew just a single food groove (not five), wound about the skeleton in a screwlike spiral. None of these groups survived the Paleozoic, and all modern echinoderms occupy the restricted realm of five-part symmetry. Yet none of these ancient groups shows any sign of anatomical insufficiency, or any hint of elimination by competition from surviving designs. Similar patterns may be found in the history of mollusks and vertebrates (where the early jawless and primitively jawed "fishes" show more variation in number and order of bones than all the later birds, reptiles, and mammals could muster; outward variety based on stereotypy of anatomical design has become a vertebrate hallmark).*

In my recent studies I concluded that the pattern of maximal early breadth is a general characteristic of lineages at several scales and times, not only of major groups at the Cambrian explosion. In fact, we have proposed that this "bottom-heavy" asymmetry may rank among the few natural phenomena imparting a direction to time, thus serving as a rare example of "time's arrow" (Gould, Gilinsky, and German, 1987; Morris,

---

*The repetition of the Burgess pattern by conventional groups with hard parts is very fortunate and favorable for testing the main issue presented by the phenomenon of decimation: Do losers disappear by inferiority in competition, or by lottery? Unfortunately, we can learn little about this key question from the Burgess Shale itself, for this soft-bodied fauna is only a spot in time, and we have virtually no evidence about the pattern of later decimation. (One Devonian arthropod, *Mimetaster* from the Hunsrückschiefer, is probably a surviving relative of *Marrella;* most other Burgess anatomies disappear without issue, and we have no evidence at all for how or when.) But patterns of extinction in groups with hard parts can be traced. Paradoxically, therefore, the best and most operational way to test for sources of decimation in the Burgess would be to study the parallel and tractable situation in echinoderms. My first question: do echinoderm "failures" tend to disappear at full abundance during mass extinctions, or to peter out gradually at different uncoordinated times? The former situation would be strong evidence for a substantial component of lottery in decimation. We do not know the answer to this question, but the solution is obtainable in principle.

1984). In our study, we portrayed evolutionary lineages and taxonomic groups as the traditional "spindle diagrams" of paleontology—read intuitively with the vertical dimension as time, and the width at any time proportional to the number of representatives in the group then living (figure 5.4). These diagrams may be bottom-heavy, top-heavy, or symmetrical (with maximum representation in the middle of the geological range). If bottom-heavy lineages characterize the history of life, then the Burgess pattern has generality across scales (for most of our spindle diagrams portray groups of low taxonomic rank, usually genera within families). If symmetrical lineages predominate, then the shape of diversification gives no direction to time.

We measure degree of asymmetry by the relative position of the diagram's center of gravity. This statement may sound like a mouthful, but our measure is intuitive and easy to grasp. Lineages with centers of gravity less than 0.5 (bottom-heavy in our terminology) reach their greatest diversity before their halfway point—that is, they follow the Burgess pattern. Lineages with centers of gravity above 0.5 attain their greatest representation past the halfway point of their geological lifetimes (see figure 5.4).

In this way, we surveyed the entire history of marine invertebrate life—

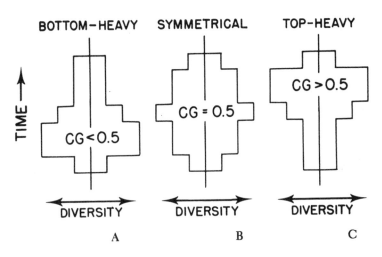

5.4. Centers of gravity in paleontological spindle diagrams. (A) A bottom-heavy diagram, with center of gravity less than 0.5. (B) A symmetrical diagram, with center of gravity at 0.5. (C) A top-heavy diagram, with center of gravity greater than 0.5.

708 separate spindle diagrams at the level of genera within families. We found only one pattern of statistically significant departure from symmetry. Lineages that arose early in the history of multicellular life, defined as during the Cambrian or Ordovician periods, have average centers of gravity less than 0.5. Lineages that arose later cannot be distinguished from 0.5 in their mean values. The Burgess pattern is therefore affirmed across all groups of the conventional fossil record for marine invertebrates with hard parts. The early history of multicellular life is marked by a bottom-heavy signature for individual lineages; later times feature symmetrical lineages.

Moreover, we found the same pattern as a generality for groups in early phases of expansion. The bottom-heavy signature is not an oddity of Cambrian invertebrate life, but a general statement about the nature of evolutionary diversification. For example, mammalian lineages that arose during the Paleocene epoch, the initial period of explosive diversification following the demise of dinosaurs, tend to be bottom-heavy, while lineages arising later are symmetrical.

We may interpret this bottom-heavy pattern in several ways. I like to think of it as "early experimentation and later standardization." Major lineages seem able to generate remarkable disparity of anatomical design at the outset of their history—early experimentation. Few of these designs survive an initial decimation, and later diversification occurs only within the restricted anatomical boundaries of these survivors—later standardization. The number of species may continue to increase, and may reach maximal values late in the history of lineages, but these profound diversifications occur within restricted anatomies—nearly a million described species of modern insects, but only three basic arthropod designs today, compared with more than twenty in the Burgess.

However we interpret this bottom-heavy pattern, it strongly reinforces the case for contingency, and validates the principal theme of this book. First, the basic pattern is a disproof of our standard and comfortable iconography—the cone of increasing diversity. The thrall of this iconography and its underlying conceptual base prevented Walcott from grasping the true extent of Burgess disparity, and has continued to portray the controlling pattern of evolution in a direction opposite to its actual form. Second, maximal initial disparity and later decimation give the broadest possible role to contingency, for if the current taxonomic structure of life records the few fortunate survivors in a lottery of decimation, rather than the end result of progressive diversification by adaptive improvement, then a replay of life's tape would yield a substantially different set of surviving anatomies and a later history making perfect sense in its own terms but markedly different from the one we know.

MASS EXTINCTION

If we could move continuously from the small to the large in inferring the causes of evolution, then Darwinian processes of the here and now might construct the topologies of evolutionary trees by extension. Since Darwin himself read a message of progress, albeit fitfully and ambiguously, in this theme of extrapolation from small to large, any geologically based derailment of this accumulative model would remove the best available argument for predictable advance in the history of life.

Mass extinctions have been recorded since the dawn of paleontology. These episodes mark the major boundaries of the geological time scale. Yet, two aspects of Darwinian tradition have led paleontologists, until the last decade, to incorporate mass extinctions into the accumulative model. First, one could try, as Darwin himself did, to portray mass extinctions as artifacts of an imperfect fossil record. Rates of dying may have been genuinely high in these times, but the extinctions were probably spread rather evenly over several million years, and only have the appearance of geological simultaneity because most times are not represented by any sediment, and the extended period of extinction may be compressed into a single bedding plane. Second, one could grant that such episodes were especially rapid, but argue that the enhanced stress only "turns up the gain" on Darwinian processes slated to yield progress: if competition in ordinary times gradually precipitates out the best, just think what the incomparably fiercer battles in an immeasurably tougher world might produce. Mass extinction should only accelerate the process of predictable advance.

The subject of mass extinction has received a new life in excitement, novel ideas, and hard data during the past ten years. The initial stimulant was, of course, Alvarez's theory of extinction triggered by extraterrestrial impact, but the discussion has moved well beyond errant asteroids to comet showers, putative 26-million-year cycles, and mathematical models for genuine catastrophe. An adequate account of this work would take a book in itself, but I do discern a general theme that can be epitomized in a statement with far-ranging implications: mass extinctions are *more frequent, rapid, devastating in magnitude,* and *distinctively different in effect* than we formerly imagined. Mass extinctions, in other words, seem to be genuine disruptions in geological flow, not merely the high points of a continuity. They may result from environmental change at such a rate, and with so drastic a result, that organisms cannot adjust by the usual forces of natural selection. Thus, mass extinctions can derail, undo, and reorient whatever might be accumulating during the "normal" times between.

The main question raised by mass extinction has always been, Is there any pattern to who gets through and who doesn't—and if so, what causes the pattern? The most exciting prospect raised by new views on mass extinction holds that the reasons for differential survival are qualitatively different from the causes of success in normal times—thus imparting a distinctive, and perhaps controlling, signature to diversity and disparity in the history of life. Such a distinctively geological, large-scale agent of pattern would disprove the old accumulative model that offered to the doctrine of progress its best remaining hope. Paleontologists are just beginning to study the causal structure of differential survival, and the jury will be out for some time. But we already have strong indications that two models of patterning by mass extinction—I call them the random and the different-rules models—not only make the case for distinctiveness but also greatly strengthen the theme of contingency.

1. *The random model.* I need hardly say that if a mass extinction operates like a genuine lottery, with each group holding a ticket unrelated to its anatomical virtues, then contingency, and maximal range of possibilities in replaying life's tape, have been proven. We have some indications that true randomness may play a role. Some of the events are so profound, and the pool of survivors so restricted, that chance fluctuations in small samples may come into play. David M. Raup, for example, has estimated species loss in the Permo-Triassic extinction, the granddaddy of all, at 96 percent. When diversity plummets to 4 percent of its former value, we must entertain the idea that some groups lose by something akin to sheer bad luck.

In a more direct study, Jablonski (1986) has traced the role in mass extinction of features known either to promote survival or to enhance speciation for marine mollusks in normal times. Jablonski found that none of these factors was beneficial or detrimental to survival in the different conditions of a mass extinction. With respect, at least, to these important causal factors of normal times, mass extinctions preserve or annihilate species at random. Geographic range was about the only factor that Jablonski could correlate with probability of survival—the bigger the area inhabited by a group, the greater its chance of pulling through. Perhaps times are so tough at these moments that the more space you normally occupy, the better your chance of finding someplace to hide.*

2. *The different-rules model.* I don't, myself, believe that true randomness predominates in mass extinctions (though it probably plays some role,

---

*Geographic range is a property of populations, not of individual clams or snails. Hence, even if survival is correlated with geographic range, a species' fate may be random with respect to the anatomical virtues of its individuals.

particularly in the most profound of the great dyings). I think that most survivors get through for specific reasons, often a complex set of causes. But I also strongly suspect that in a great majority of cases, the traits that enhance survival during an extinction do so in ways that are incidental and unrelated to the causes of their evolution in the first place.

This contention is the centerpiece of the different-rules model. Animals evolve their sizes, shapes, and physiologies under natural selection in normal times, and for specifiable reasons (usually involving adaptive advantage). Along comes a mass extinction, with its "different rules" for survival. Under the new regulations, the very best of your traits, the source of your previous flourishing, may now be your death knell. A trait with no previous significance, one that had just hitchhiked along for the developmental ride as a side consequence of another adaptation, may now hold the key to your survival. There can be no causal correlation in principle between the reasons for evolving a feature and its role in survival under the new rules. (The key issue for testing this model therefore lies in establishing that new rules do, indeed, prevail.) A species, after all, cannot evolve structures with a view to their potential usefulness millions of years down the road—unless our general ideas about causality are markedly awry, and the future can control the present.

We probably owe our own existence to such good fortune. Small animals, for reasons not well understood, seem to have an edge in most mass extinctions, particularly in the Cretaceous event that wiped out remaining dinosaurs. Mammals may therefore have survived that great dying primarily because they were small, not because they embodied any intrinsic anatomical virtues relative to dinosaurs, now doomed by their size. And mammals were surely not small because they had sensed some future advantage; they had probably remained small for a reason that would be judged negatively in normal times—because dinosaurs dominated environments for large terrestrial vertebrates, and incumbents have advantages in nature as well as in politics.

Kitchell, Clark, and Gombos (1986) have worked out an interesting example based on diatoms, single-celled plants of the oceanic plankton. Paleontologists have long wondered why diatoms came through the Cretaceous extinction relatively unscathed, while most other elements of the plankton crashed. For growth and reproduction, diatoms rely upon the seasonal availability of nutrients rising to the surface from deeper waters in zones of upwelling. (These episodes of upwelling unleash so-called diatom "blooms.") When these nutrients are depleted, diatoms can change their form to a "resting spore," essentially shut down their metabolism and sink to deeper waters. A return of nutrients will terminate this period of dormancy. Kitchell and her colleagues attribute the success of diatoms in the Cretaceous extinction to an incidental side consequence of dormancy. The resting spores evolved as a strategy for dealing with predictable and seasonal fluctuations in nutrients, clearly not for environmental catastrophes of mass extinction. But the ability to hunker down in a dormant state may have saved the diatoms under the different rules of mass extinction, especially if the "nuclear winter" model proves valid for the Cretaceous event—for darkness would cut off photosynthesis and propagate extinctions up and down a food chain ultimately dependent upon primary production, while diatoms might ride out the dark storm as resting spores below the photic zone.

The different-rules model therefore fractures the causal continuity that Darwin envisaged between reasons for success within local populations and the causes of survival and proliferation through long stretches of geological time. Hence, this model strongly promotes the role of contingency, viewed primarily as unpredictability, in evolution. If long-term success depends upon incidental aspects of features evolved for different reasons, then how could we possibly know, if we rewound life's tape to a distant past, which groups were destined for success? Their performance and evolution during our observation would not be relevant. We might base some guesses on incidental features that usually imply survival through a mass extinction, but how could we do so with any confidence? In an important sense, these crucial features don't even exist until the different rules of mass extinction make their incidental effects important—for extreme stress may be needed to "key up" these features, and animals may never experience such conditions during normal times. And how can we know, in our rich and multifarious world, what the next episode of mass extinction, somewhere down the road, will require? Unpredictability must rule if geological longevity depends upon lucky side consequences of features evolved for other reasons.

I particularly welcome this demonstration that several general principles of large-scale evolution promote the importance of contingency. The generalizations—on the bottom-heaviness of lineages and the properties of mass extinctions—are the stuff of traditional nonhistorical science, the style that usually opposes, or at least downgrades, a historical principle like contingency. This reinforcement is a happy situation for scientific pluralism. I do not relish the idea of defending historical science by building a bunker and fighting for respect and self-determination. Better to move forward in partnership; general patterns of evolution imply the unpredictability of specific outcomes.

SEVEN POSSIBLE WORLDS

The collapse of the cone and the ladder opens the floodgates to alternative worlds that didn't emerge, but might have arisen with slight and sensible changes in some early events. These unrealized universes would have been every bit as ordered and explainable as the world we know, but ever so different in ways that we can never specify in detail. The enumeration of unrealized worlds is a parlor game without end, for who can count the possibilities? The universe is not so tightly interconnected that the fall of a petal disrupts a distant star, whatever our poets sing. But most quirky changes of topography or environment, most appearances and disappearances of groups (if not of single species), can irrevocably alter the pathways of life in substantial ways. The playground of contingency is immeasurable. Let us consider just seven alternative scenarios, arranged in chronological order to home in on the biological object that most excites our parochial fancy—*Homo sapiens.*

EVOLUTION OF THE EUKARYOTIC CELL

Life arose at least 3.5 billion years ago, about as soon as the earth became cool enough for stability of the chief chemical components. (I do not, by the way, view the origin of life itself as a chancy or unpredictable event. I suspect that given the composition of early atmospheres and oceans, life's origin was a chemical necessity. Contingency arises later, when historical complexity enters the picture of evolution.)

With respect to the old belief in steady progress, nothing could be stranger than the early evolution of life—for nothing much happened for ever so long. The oldest fossils are prokaryotic cells some 3.5 billion years old (see pages 57–58). The fossil record of this time also includes the highest form of macroscopic complexity evolved by these prokaryotes—stromatolites. These are layers of sediment trapped and bound by prokaryotic cells. The layers may pile up one atop the other, as tides bury and re-form the mats—and the whole structure may come to resemble a cabbage in cross section (also in size).

Stromatolites and their prokaryotic builders dominated the fossil record throughout the world for more than 2 billion years. The first eukaryotic cells (the complex textbook variety, complete with nucleus and numerous structures of the cytoplasm) appeared some 1.4 billion years ago. The conventional argument holds that eukaryotic cells are a prerequisite for multicellular complexity, if only because sexual reproduction required paired chromosomes, and only sex can supply the variation that natural selection needs as raw material for further complexity.

But multicellular animals did not arise soon after the origin of eukaryotic cells; they first appeared just before the Cambrian explosion some 570 million years ago. Hence, a good deal more than half the history of life is a story of prokaryotic cells alone, and only the last one-sixth of life's time on earth has included multicellular animals.

Such delays and long lead times strongly suggest contingency and a vast realm of unrealized possibilities. If prokaryotes had to advance toward eukaryotic complexity, they certainly took their time about it. Moreover, when we consider the favored hypothesis for the origin of the eukaryotic cell, we enter the realm of quirky and incidental side consequences as unpredictable sources of change. Our best theory identifies at least some major organelles—the mitochondria and chloroplasts almost surely, and others with less confidence—as descendants of entire prokaryotic cells that evolved to live symbiotically within other cells (Margulis, 1981). In this view, each eukaryotic cell is, by descent, a colony that later achieved tighter integration. Surely, the mitochondrion that first entered another cell was not thinking about the future benefits of cooperation and integration; it was merely trying to make its own living in a tough Darwinian world. Accordingly, this fundamental step in the evolution of multicellular life arose for an immediate reason quite unrelated to its eventual effect upon organic complexity. This scenario seems to portray fortunate contingency rather than predictable cause and effect. And if you wish nevertheless to view the origin of organelles and the transition from symbiosis to integration as predictable in some orderly fashion, then tell me why more than half the history of life passed before the process got started.

One final point that I find chilling with respect to the possibility of something like human evolution in an alternative world: Even though this first event took more than half the known history of life, I might be prepared to accept the probability of an eventual origin for higher intelligence if the earth were slated to endure for hundreds of billion of years—so that this initial step took but a tiny fraction of potential time. But cosmologists tell us that the sun is just about at the halfway point of existence in its current state; and that some five billion years from now, it will explode, expanding in diameter beyond the orbit of Jupiter and engulfing the earth. Life will end unless it can move elsewhere; and life on earth will terminate in any case.

Since human intelligence arose just a geological second ago, we face the stunning fact that the evolution of self-consciousness required about half of the earth's potential time. Given the errors and uncertainties, the variations of rates and pathways in other runs of the tape, what possible confidence can we have in the eventual origin of our distinctive mental abilities? Run the tape again, and even if the same general pathways emerge, it might take twenty billion years to reach self-consciousness this time—except that the earth would be incinerated billions of years before. Run the tape again, and the first step from prokaryotic to eukaryotic cell might take twelve billion instead of two billion years—and stromatolites, never awarded the time needed to move on, might be the highest mute witnesses to Armageddon.

THE FIRST FAUNA OF MULTICELLULAR ANIMALS

You might accept this last sobering scenario, but then claim, fine, I'll grant the unpredictability of getting beyond prokaryotic cells, but once you finally do get multicellular animals, then the basic pathways are surely set and further advance to consciousness must occur. But let's take a closer look.

The first multicellular animals, as discussed in chapter II, are members of a world-wide fauna named for the most famous outcrop at Ediacara, in Australia. Martin Glaessner, the paleontologist most responsible for describing the Ediacara animals, has always interpreted them, under traditional concepts of the cone, as primitive representatives of modern groups—mostly members of the coelenterate phylum (soft corals and medusoids), but including annelid worms and arthropods (Glaessner, 1984). Glaessner's traditional reading evoked very little opposition (but see Pflug, 1972 and 1974), and the Ediacara fauna settled comfortably into textbooks as fitting ancestors for modern groups—for their combination of maximal age with minimal complexity neatly matches expectations.

The Ediacara fauna has special importance as the only evidence for multicellular life before the great divide separating the Precambrian and Cambrian, a boundary marked by the celebrated Cambrian explosion of modern groups with hard parts. True, the Ediacara creatures are only barely Precambrian; they occur in strata just predating Cambrian and probably do not extend more than 100 million years into the uppermost Precambrian. In keeping with their position immediately below the boundary, the Ediacara animals are entirely soft-bodied. If taxonomic identity could be maintained right through this greatest of geological transitions, and without major disruption in design to accompany the evolution of hard parts, then the smooth continuity of the cone would be confirmed. This version of Ediacara begins to sound suspiciously like Walcott's shoehorn.

In the early 1980s, my friend Dolf Seilacher, professor of paleontology at Tübingen, Germany, and in my opinion the finest paleontological observer now active, proposed a radically different interpretation of the Ediacara fauna (Seilacher, 1984). His twofold defense rests upon a negative and a positive argument. For his negative claim, he argues on functional grounds that the Ediacara creatures could not have operated as their supposed modern counterparts, and therefore may not be allied with any living group, despite some superficial similarity of outward form. For example, most Ediacara animals have been allied with the soft corals, a group including the modern sea fans. Coral skeletons represent colonies housing thousands of tiny individuals. In soft corals, the individual polyps line the branches of a tree or network structure, and the branches must be separated, so that water can bring food particles to the polyps and sweep away waste products. But the apparent branches of the Ediacara forms are joined together, forming a flattened quiltlike mat with no spaces between the sections.

For his positive claim, Seilacher argues that most Ediacara animals may be taxonomically united as variations on a single anatomical plan—a flattened form divided into sections that are matted or quilted together, perhaps constituting a hydraulic skeleton much like an air mattress (figure 5.5). Since this design matches no modern anatomical plan, Seilacher concludes that the Ediacara creatures represent an entirely separate experiment in multicellular life—one that ultimately failed in a previously unrecognized latest Precambrian extinction, for no Ediacara elements survived into the Cambrian.

For the Burgess fauna, the case against Walcott's shoehorn has been proven, I think, with as much confidence as science can muster. For the Ediacara fauna, Seilacher's hypothesis is a plausible and exciting, but as yet unproven, alternative to the traditional reading, which will one day be called either Glaessner's shoehorn or Glaessner's insight, as the case may be.

But consider the implications for unpredictability if Seilacher's view prevails, even partly. Under Glaessner's ranking in modern groups, the first animals share the anatomical designs of later organisms, but in simpler form—and evolution must be channeled up and outward in the traditional cone of increasing diversity. Replay the tape, starting with simple coelenterates, worms, and arthropods, a hundred times, and I suppose that you will usually end up with more and better of the same.

But if Seilacher is right, other possibilities and other directions were once available. Seilacher does not believe that all late Precambrian animals fall within the taxonomic boundaries of this alternative and independent

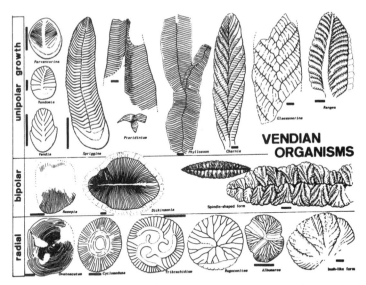

5.5. Seilacher's classification of the Ediacara organisms according to their variations on a single flattened, quiltlike anatomical plan. These organisms are conventionally placed in several different modern phyla.

experiment in multicellular life. By studying the varied and abundant trace fossils (tracks, trails, and burrows) of the same strata, he is convinced that metazoan animals of modern design—probably genuine worms in one form or another—shared the earth with the Ediacara fauna. Thus, as with the Burgess, several different anatomical possibilities were present right at the beginning. Life might have taken either the Ediacara or the modern pathway, but Ediacara lost entirely, and we don't know why.

Suppose that we could replay life's tape from late Precambrian times, and that the flat quilts of Ediacara won on their second attempt, while metazoans were eliminated. Could life have ever moved to consciousness along this alternate pathway of Ediacara anatomy? Probably not. Ediacara design looks like an alternative solution to the problem of gaining enough surface area as size increases. Since surfaces (length^2) increase so much more slowly than volumes (length^3), and since animals perform most functions through surfaces, some way must be found to elaborate surface area in large creatures. Modern life followed the path of evolving internal organs (lungs, villi of the small intestine) to provide the requisite surfaces. In a second solution—proposed by Seilacher as the key to understanding Ediacara design—organisms may not be able to evolve internal complexity and must rely instead on changes in overall form, taking the shape of threads, ribbons, sheets, or pancakes so that no internal space lies very far from the outer surface. (The complex quilting of Ediacara animals could then be viewed as a device for strengthening such a precarious form. A sheet one foot long and a fraction of an inch thick needs some extra support in a world of woe, tides, and storms.)

If Ediacara represents this second solution, and if Ediacara had won the replay, then I doubt that animal life would ever have gained much complexity, or attained anything close to self-consciousness. The developmental program of Ediacara creatures might have foreclosed the evolution of internal organs, and animal life would then have remained permanently in the rut of sheets and pancakes—a most unpropitious shape for self-conscious complexity as we know it. If, on the other hand, Ediacara survivors had been able to evolve internal complexity later on, then the pathways from this radically different starting point would have produced a world worthy of science fiction at its best.

THE FIRST FAUNA OF THE CAMBRIAN EXPLOSION

Our hypothetical advocate of the cone and ladder might be willing to give ground on these first two incidents from the dim mists of time, but he might then be tempted to dig his entrenchment across the Cambrian boundary. Surely, once the great explosion occurs, and traditional fossils with hard parts enter the record, then the outlines must be set, and life must move upward and outward in predictable channels.

Not so. As noted in chapter II, the initial shelly fauna, called Tommotian to honor a famous Russian locality, contains far more mysteries than precursors. Some modern groups make an undoubted first appearance in the Tommotian, but more of these fossils may represent anatomies beyond the current range. The story is becoming familiar—a maximum of potential pathways at the beginning, followed by decimation to set the modern pattern.

The most characteristic and abundant of all Tommotian creatures, the archaeocyathids (figure 5.6), represent a long-standing problem in classification. The familiar litany plays again. These first reef-forming creatures of the fossil record are simple in form, usually cone-shaped, with double walls—cup within cup. In the traditional spirit of the shoehorn, they have been shunted from one modern group to another during more than a century of paleontological speculation. Corals and sponges have been their usual putative homes. But the more we learn about archaeocyathids, the stranger they appear, and most paleontologists now place them in a sepa-

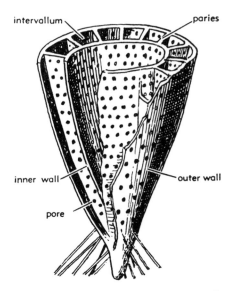

5.6. An archaeocyathid, showing the basic organization of cup within cup.

rate phylum destined to disappear before the Cambrian had run its course.

Even more impressive is the extensive disparity just now being recognized among organisms of the "small shelly fauna." Tommotian rocks house an enormous variety of tiny fossils (usually one to five millimeters in length) that cannot be allied with any modern group (Bengtson, 1977; Bengtson and Fletcher, 1983). We can arrange these fossils by outward appearance, as tubes, spines, cones, and plates (figure 5.7 shows a representative sample), but we do not know their zoological affinities. Perhaps they are merely bits and pieces from an era of early, still imperfect skeletonization; perhaps they covered familiar organisms that later developed the more elaborate shells of their conventional fossil signatures. But perhaps—and this interpretation has recently been gaining favor among aficionados of the small shelly fauna—most of the Tommotian oddballs represent unique anatomies that arose early and disappeared quickly. For example, Rozanov, the leading Russian expert on this fauna, concludes his recent review by writing:

> Early Cambrian rocks contain numerous remains of very peculiar organisms, both animals and plants, most of which are unknown after the Cambrian. I tend to think that numerous high-level taxa developed in the early Cambrian and rapidly became extinct (1986, p. 95).

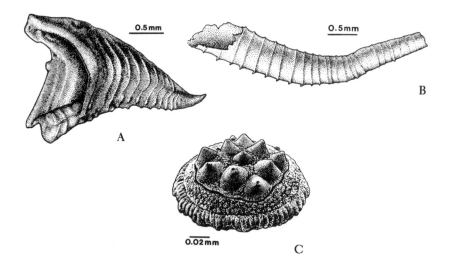

5.7. Representative organisms of unknown affinity from the Cambrian "small shelly fauna" (Rozanov, 1986). (A) *Tommotia.* (B) *Hyolithellus.* (C) *Lenargyrion.*

Once again, we have a Christmas tree rather than a cone. Once again, the unpredictability of evolutionary pathways asserts itself against our hope for the inevitability of consciousness. The Tommotian contained many modern groups, but also a large range of alternative possibilities. Rewind the tape into the early Cambrian, and perhaps this time our modern reefs are built by archaeocyathids, not corals. Perhaps no Bikini, no Waikiki; perhaps, also, no people to sip rum swizzles and snorkle amidst great undersea gardens.

THE SUBSEQUENT CAMBRIAN ORIGIN OF THE MODERN FAUNA

Our traditionalist is now beginning to worry, but he will grant this one last point *pour mieux sauter.* OK, the very first Cambrian fauna included a plethora of alternative possibilities, all equally sensible and none leading to us. But, surely, once the modern fauna arose in the next phase of the Cambrian, called Atdabanian after another Russian locality, then the boundaries and channels were finally set. The arrival of trilobites, those familiar symbols of the Cambrian, must mark the end of craziness and the inception of predictability. Let the good times roll.

This book is quite long enough already, and you do not want a "second verse, same as the first." I merely point out that the Burgess Shale represents the early and maximal extent of the Atdabanian radiation. The story of the Burgess Shale is the tale of life itself, not a unique and peculiar episode of possibilities gone wild.

THE ORIGIN OF TERRESTRIAL VERTEBRATES

Our traditionalist is now reeling. He is ready to abandon virtually all of life to contingency, but he will make his last stand with vertebrates. The game, after all, centers on human consciousness as the unpredictable product of an incidental twig, or the culmination of an ineluctable, or at the very least a probable, trend. To hell with the rest of life; they aren't on the lineage leading to consciousness in any case. Surely, once vertebrates arose, however improbable their origin, we could then mount confidently from ponds to dry land to hind legs to big brains.

I might grant the probability of the most crucial environmental transition—from water to land—if the characteristic anatomy of fishes implied, even for incidental reasons, an easy transformation of fins into sturdy limbs needed for support in the gravity of terrestrial environments. But the fins of most fishes are entirely unsuited for such a transition. A stout basal bar follows the line of the body axis, and numerous thin fin rays run parallel to each other and perpendicular to the bar. These thin, unconnected rays could not support the weight of the body on land. The few modern fishes that scurry across mud flats, including *Periophthalmus,* the "walking fish," pull their bodies along and do not stride with their fins.

Terrestrial vertebrates could arise because a relatively small group of fishes, only distantly related to the "standard issue," happened, for their own immediate reasons, to evolve a radically different type of limb skeleton, with a strong central axis perpendicular to the body, and numerous lateral branches radiating from this common focus. A structure of this design could evolve into a weight-bearing terrestrial limb, with the central axis converted to the major bones of our arms and legs, and the lateral branches forming digits. Such a fin structure did not evolve for its future flexibility in permitting later mammalian life; (this limb may have provided advantages, in superior rotation, for bottom-dwelling fishes that used the substrate as an aid in propulsion). But whatever its unknown advantages, this necessary prerequisite to terrestrial life evolved in a restricted group of fishes off the main line—the lungfish-coelacanth-rhipidistian complex. Wind the tape of life back to the Devonian, the so-called age of fishes. Would an observer have singled out these uncommon and uncharacteristic fishes as precursors to such conspicuous success in such a different environment? Replay the tape, expunge the rhipidistians by extinction, and our lands become the unchallenged domain of insects and flowers.

PASSING THE TORCH TO MAMMALS

Can we not grant the traditionalist some solace? Let contingency rule right to the origin of mammals. Can we not survey the world as mammals emerged into the realm of dinosaurs, and know that the meek and hairy would soon inherit the earth? What defense could large, lumbering, stupid, cold-blooded behemoths provide against smarts, sleekness, live birth, and constant body temperature? Don't we all know that mammals arose late in the reign of dinosaurs; and did they not then hasten the inevitable transition by eating their rivals' eggs?

This common scenario is fiction rooted in traditional hopes for progress and predictability. Mammals evolved at the end of the Triassic, at the same time as dinosaurs, or just a tad later. Mammals spent their first hundred million years—two-thirds of their total history—as small creatures living in the nooks and crannies of a dinosaur's world. Their sixty million years of success following the demise of dinosaurs has been something of an afterthought.

We have no indication of any trend toward mammalian hegemony during this initial hundred million years. Quite the reverse—dinosaurs remained in unchallenged possession of all environments for large-bodied terrestrial creatures. Mammals made no substantial moves toward domination, larger brains, or even greater size.

If mammals had arisen late and helped to drive dinosaurs to their doom, then we could legitimately propose a scenario of expected progress. But dinosaurs remained dominant and probably became extinct only as a quirky result of the most unpredictable of all events—a mass dying triggered by extraterrestrial impact. If dinosaurs had not died in this event, they would probably still dominate the domain of large-bodied vertebrates, as they had for so long with such conspicuous success, and mammals would still be small creatures in the interstices of their world. This situation prevailed for a hundred million years; why not for sixty million more? Since dinosaurs were not moving toward markedly larger brains, and since such a prospect may lie outside the capabilities of reptilian design (Jerison, 1973; Hopson, 1977), we must assume that consciousness would not have evolved on our planet if a cosmic catastrophe had not claimed the dinosaurs as victims. In an entirely literal sense, we owe our existence, as large and reasoning mammals, to our lucky stars.

THE ORIGIN OF *Homo sapiens*

I will not carry this argument to ridiculous extremes. Even I will admit that at some point in the story of human evolution, circumstances conspired to encourage mentality at our modern level. The usual scenario holds that attainment of upright posture freed the hands for using tools and weapons, and feedback from the behavioral possibilities thus provided spurred the evolution of a larger brain.

But I believe that most of us labor under a false impression about the pattern of human evolution. We view our rise as a kind of global process encompassing all members of the human lineage, wherever they may have lived. We recognize that *Homo erectus,* our immediate ancestor, was the first species to emigrate from Africa and to settle in Europe and Asia as well ("Java Man" and "Peking Man" of the old texts). But we then revert to the hypothesis of global impetus and imagine that all *Homo erectus* populations on all three continents moved together up the ladder of mentality on a wave of predictable and necessary advance, given the adaptive value of intelligence. I call this scenario the "tendency theory" of human evolution. *Homo sapiens* becomes the anticipated result of an evolutionary tendency pervading all human populations.

In an alternative view, recently given powerful support by reconstructions of our evolutionary tree based on genetic differences among modern groups (Cann, Stoneking, and Wilson, 1987; Gould, 1987b), *Homo sapiens* arose as an evolutionary item, a definite entity, a small and coherent population that split off from a lineage of ancestors in Africa. I call this view the "entity theory" of human evolution. It carries a cascade of arresting implications: Asian *Homo erectus* died without issue and does not enter our immediate ancestry (for we evolved from African populations); Neanderthal people were collateral cousins, perhaps already living in Europe while we emerged in Africa, and also contributing nothing to our immediate genetic heritage. In other words, we are an improbable and fragile entity, fortunately successful after precarious beginnings as a small population in Africa, not the predictable end result of a global tendency. We are a thing, an item of history, not an embodiment of general principles.

This claim would not carry startling implications if we were a repeatable thing—if, had *Homo sapiens* failed and succumbed to early extinction as most species do, another population with higher intelligence in the same form was bound to originate. Wouldn't the Neanderthals have taken up the torch if we had failed, or wouldn't some other embodiment of mentality at our level have originated without much delay? I don't see why. Our closest ancestors and cousins, *Homo erectus*, the Neanderthals, and others, possessed mental abilities of a high order, as indicated by their range of tools and other artifacts. But only *Homo sapiens* shows direct evidence for the kind of abstract reasoning, including numerical and aesthetic modes, that we identify as distinctively human. All indications of ice-age reckoning—the calendar sticks and counting blades—belong to *Homo sapiens*. And all the ice-age art—the cave paintings, the Venus figures, the horsehead carvings, the reindeer bas-reliefs—was done by our species. By evidence now available, Neanderthal knew nothing of representational art.

Run the tape again, and let the tiny twig of *Homo sapiens* expire in Africa. Other hominids may have stood on the threshold of what we know as human possibilities, but many sensible scenarios would never generate our level of mentality. Run the tape again, and this time Neanderthal perishes in Europe and *Homo erectus* in Asia (as they did in our world). The sole surviving human stock, *Homo erectus* in Africa, stumbles along for a while, even prospers, but does not speciate and therefore remains stable. A mutated virus then wipes *Homo erectus* out, or a change in climate reconverts Africa into inhospitable forest. One little twig on the mammalian branch, a lineage with interesting possibilities that were never realized, joins the vast majority of species in extinction. So what? Most possibilities are never realized, and who will ever know the difference?

Arguments of this form lead me to the conclusion that biology's most profound insight into human nature, status, and potential lies in the simple phrase, the embodiment of contingency: *Homo sapiens* is an entity, not a tendency.

By taking this form of argument across all scales of time and extent, and right to the heart of our own evolution, I hope I have convinced you that contingency matters where it counts most. Otherwise, you may view this projected replaying of life's tape as merely a game about alien creatures. You may ask if all my reveries really make any difference. Who cares, in the old spirit of America at its pragmatic best? It is fun to imagine oneself as a sort of divine disk jockey, sitting before the tape machine of time with a library of cassettes labeled "priapulids," "polychaetes," and "primates." But would it really matter if all the replays of the Burgess Shale produced their unrealized opposites—and we inhabited a world of wiwaxiids, a sea floor littered with little penis worms, and forests full of phororhacids? We might be shucking sclerites instead of opening shells for our clambakes. Our trophy rooms might vie for the longest *Diatryma* beak, not the richest lion mane. But what would be fundamentally different?

Everything, I suggest. The divine tape player holds a million scenarios, each perfectly sensible. Little quirks at the outset, occurring for no particular reason, unleash cascades of consequences that make a particular future seem inevitable in retrospect. But the slightest early nudge contacts a different groove, and history veers into another plausible channel, diverging continually from its original pathway. The end results are so different, the initial perturbation so apparently trivial. If little penis worms ruled the sea, I have no confidence that *Australopithecus* would ever have walked erect on the savannas of Africa. And so, for ourselves, I think we can only exclaim, O brave—and improbable—new world, that has such people in it!

An Epilogue on *PIKAIA*

I must end this book with a confession. I pulled a small, and I trust harmless, pedagogical trick on you. In my long discussion of Burgess Shale organisms, I purposely left one creature out. I might offer the flimsy excuse that Simon Conway Morris has not yet published his monograph on this genus—for he has been saving the best for last. But that claim would be disingenuous. I forbore because I also wanted to save the best for last.

In his 1911 paper on supposed Burgess annelids, Walcott described an attractive species, a laterally compressed ribbon-shaped creature some two inches in length (figure 5.8). He named it *Pikaia gracilens,* to honor nearby Mount Pika, and to indicate a certain elegance of form. Walcott confidently placed *Pikaia* among the polychaete worms. He based this classification on the obvious and regular segmentation of the body.

Simon Conway Morris therefore received *Pikaia* along with his general thesis assignment of the Burgess "worms." As he studied the thirty or so specimens of *Pikaia* then known, he reached a firm conclusion that others had suspected, and that had circulated around the paleontological rumor mills for some time. *Pikaia* is not an annelid worm. It is a chordate, a member of our own phylum—in fact, the first recorded member of our immediate ancestry. (Realizing the importance of this insight, Simon wisely saved *Pikaia* for the last of his Burgess studies. When you have something rare and significant, you must be patient and wait until your thoughts are settled and your techniques honed to their highest craft; for this is the one, above all, that you must get right.)

The structures that Walcott had identified as annelid segments exhibit the characteristic zigzag bend of chordate myotomes, or bands of muscle. Furthermore, *Pikaia* has a notochord, the stiffened dorsal rod that gives our phylum, Chordata, its name. In many respects *Pikaia* resembles, at least in general level of organization, the living *Amphioxus*—long used in

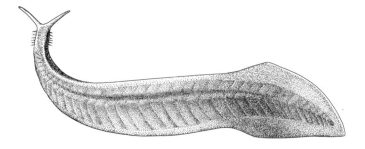

5.8. *Pikaia*, the world's first known chordate, from the Burgess Shale. Note the features of our phylum: the notochord or stiffened rod along the back that evolved into our spinal column, and the zigzag muscle bands. Drawn by Marianne Collins.

laboratories and lecture rooms as a model for the "primitive" organization of prevertebrate chordates. Conway Morris and Whittington declare:

> The conclusion that it [*Pikaia*] is not a worm but a chordate appears inescapable. The superb preservation of this Middle Cambrian organism makes it a landmark in the history of the phylum to which all the vertebrates, including man, belong (1979, p. 131).

Fossils of true vertebrates, initially represented by agnathan, or jawless, fishes, first appear in the Middle Ordovician, with fragmentary material of uncertain affinity from the Lower Ordovician and even the Upper Cambrian—all considerably later than the Burgess *Pikaia* (see Gagnier, Blieck, and Rodrigo, 1986).

I do not, of course, claim that *Pikaia* itself is the actual ancestor of vertebrates, nor would I be foolish enough to state that all opportunity for a chordate future resided with *Pikaia* in the Middle Cambrian; other chordates, as yet undiscovered, must have inhabited Cambrian seas. But I suspect, from the rarity of *Pikaia* in the Burgess and the absence of chordates in other Lower Paleozoic *Lagerstätten*, that our phylum did not rank among the great Cambrian success stories, and that chordates faced a tenuous future in Burgess times.

Pikaia is the missing and final link in our story of contingency—the direct connection between Burgess decimation and eventual human evolution. We need no longer talk of subjects peripheral to our parochial concerns—of alternative worlds crowded with little penis worms, of marrelliform arthropods and no mosquitoes, of fearsome anomalocarids gobbling fishes. Wind the tape of life back to Burgess times, and let it play again. If *Pikaia* does not survive in the replay, we are wiped out of future history— all of us, from shark to robin to orangutan. And I don't think that any handicapper, given Burgess evidence as known today, would have granted very favorable odds for the persistence of *Pikaia*.

And so, if you wish to ask the question of the ages—why do humans exist?—a major part of the answer, touching those aspects of the issue that science can treat at all, must be: because *Pikaia* survived the Burgess decimation. This response does not cite a single law of nature; it embodies no statement about predictable evolutionary pathways, no calculation of probabilities based on general rules of anatomy or ecology. The survival of *Pikaia* was a contingency of "just history." I do not think that any "higher" answer can be given, and I cannot imagine that any resolution could be more fascinating. We are the offspring of history, and must establish our own paths in this most diverse and interesting of conceivable universes—one indifferent to our suffering, and therefore offering us maximal freedom to thrive, or to fail, in our own chosen way.

Bibliography

Aitken, J. D., and I. A. McIlreath. 1984. The Cathedral Reef escarpment, a Cambrian great wall with humble origins. *Geos: Energy Mines and Resources, Canada* 13(1):17–19.

Allison, P. A. 1988. The role of anoxia in the decay and mineralization of proteinaceous macro-fossils. *Paleobiology* 14:139–54.

Anonymous. 1987. Yoho's fossils have world significance. *Yoho National Park Highline.*

Bengtson, S. 1977. Early Cambrian button-shaped phosphatic microfossils from the Siberian platform. *Palaeontology* 20:751–62.

Bengtson, S., and T. P. Fletcher. 1983. The oldest sequence of skeletal fossils in the Lower Cambrian of southwestern Newfoundland. *Canadian Journal of Earth Sciences* 20: 525–36.

Bethell, T. 1976. Darwin's mistake. *Harper's,* February.

Briggs, D. E. G. 1976. The arthropod *Branchiocaris* n. gen., Middle Cambrian, Burgess Shale, British Columbia. *Geological Survey of Canada Bulletin* 264:1–29.

Briggs, D. E. G. 1977. Bivalved arthropods from the Cambrian Burgess Shale of British Columbia. *Palaeontology* 20:595–621.

Briggs, D. E. G. 1978. The morphology, mode of life, and affinities of *Canadaspis perfecta* (Crustacea: Phyllocarida), Middle Cambrian, Burgess Shale, British Columbia. *Philosophical Transactions of the Royal Society, London* B 281:439–87.

Briggs, D. E. G. 1979. *Anomalocaris,* the largest known Cambrian arthropod. *Palaeontology* 22:631–64.

Briggs, D. E. G. 1981a. The arthropod *Odaraia alata* Walcott, Middle Cambrian, Burgess Shale, British Columbia. *Philosophical Transactions of the Royal Society, London* B 291:541–85.

Briggs, D. E. G. 1981b. Relationships of arthropods from the Burgess Shale and other Cambrian sequences. Open File Report 81-743, U.S. Geological Survey, pp. 38–41.

Briggs, D. E. G. 1983. Affinities and early evolution of the Crustacea: The evidence of the

Cambrian fossils. In F. R. Schram (ed.), *Crustacean Phylogeny*, pp. 1–22. Rotterdam: A. A. Balkema.

Briggs, D. E. G. 1985. Les premiers arthropodes. *La Recherche* 16:340–49.

Briggs, D. E. G., E. N. K. Clarkson, and R. J. Aldridge. 1983. The conodont animal. *Lethaia* 16:1–14.

Briggs, D. E. G., and D. Collins. 1988. A Middle Cambrian chelicerate from Mount Stephen, British Columbia. *Palaeontology* 31:779–98.

Briggs, D. E. G., and S. Conway Morris. 1986. Problematica from the Middle Cambrian Burgess Shale of British Columbia. In A. Hoffman and M. H. Nitecki (eds.), *Problematic fossil taxa*, pp. 167–83. New York: Oxford University Press.

Briggs, D. E. G., and R. A. Robison. 1984. Exceptionally preserved nontrilobite arthropods and *Anomalocaris* from the Middle Cambrian of Utah. *University of Kansas Paleontological Contributions*, Paper 111.

Briggs, D. E. G., and H. B. Whittington. 1985. Modes of life of arthropods from the Burgess Shale, British Columbia. *Transactions of the Royal Society of Edinburgh* 76:149–60.

Bruton, D. L. 1981. The arthropod *Sidneyia inexpectans*, Middle Cambrian, Burgess Shale, British Columbia. *Philosophical Transactions of the Royal Society, London* B 295:619–56.

Bruton, D. L., and H. B. Whittington. 1983. *Emeraldella* and *Leanchoilia*, two arthropods from the Burgess Shale, British Columbia. *Philosophical Transactions of the Royal Society, London* B 300:553–85.

Cann, R. L., M. Stoneking, and A. C. Wilson. 1987. Mitochondrial DNA and human evolution. *Nature* 325:31–36.

Collins, D. H. 1985. A new Burgess Shale type fauna in the Middle Cambrian Stephen Formation on Mount Stephen, British Columbia. In *Annual Meeting, Geological Society of America*, p. 550.

Collins, D. H., D. E. G. Briggs, and S. Conway Morris. 1983. New Burgess Shale fossil sites reveal Middle Cambrian faunal complex. *Science* 222:163–67.

Conway Morris, S. 1976a. *Nectocaris pteryx*, a new organism from the Middle Cambrian Burgess Shale of British Columbia. *Neues Jahrbuch für Geologie und Paläontologie*, 12:705–13.

Conway Morris, S. 1976b. A new Cambrian lophophorate from the Burgess Shale of British Columbia. *Palaeontology* 19:199–222.

Conway Morris, S. 1977a. A new entoproct-like organism from the Burgess Shale of British Columbia. *Palaeontology* 20:833–45.

Conway Morris, S. 1977b. A redescription of the Middle Cambrian worm *Amiskwia sagittiformis* Walcott from the Burgess Shale of British Columbia. *Paläontologische Zeitschrift* 51:271–87.

Conway Morris, S. 1977c. A new metazoan from the Cambrian Burgess Shale, British Columbia. *Palaeontology* 20:623–40.

Conway Morris, S. 1977d. Fossil priapulid worms. In *Special papers in Palaeontology*, vol. 20. London: Palaeontological Association.

Conway Morris, S. 1978. *Laggania cambria* Walcott: A composite fossil. *Journal of Paleontology* 52:126–31.

Conway Morris, S. 1979. Middle Cambrian polychaetes from the Burgess Shale of British Columbia. *Philosophical Transactions of the Royal Society, London* B 285:227–274.

Conway Morris, S. 1985. The Middle Cambrian metazoan *Wiwaxia corrugata* (Matthew) from the Burgess Shale and *Ogygopsis* Shale, British Columbia, Canada. *Philosophical Transactions of the Royal Society, London* B 307:507–82.

Conway Morris, S. 1986. The community structure of the Middle Cambrian phyllopod bed (Burgess Shale). *Palaeontology* 29:423–67.

Conway Morris, S., J. S. Peel, A. K. Higgins, N. J. Soper, and N. C. Davis. 1987. A Burgess Shale-like fauna from the Lower Cambrian of north Greenland. *Nature* 326:181–83.

Conway Morris, S., and R. A. Robison. 1982. The enigmatic medusoid *Peytoia* and a comparison of some Cambrian biotas. *Journal of Paleontology* 56:116–22.

Conway Morris, S., and R. A. Robison. 1986. Middle Cambrian priapulids and other soft-bodied fossils from Utah and Spain. *University of Kansas Paleontological Contributions,* Paper 117.

Conway Morris, S., and H. B. Whittington. 1979. The animals of the Burgess Shale. *Scientific American* 240 (January): 122–33.

Conway Morris, S., and H. B. Whittington. 1985. Fossils of the Burgess Shale. A national treasure in Yoho National Park, British Columbia. *Geological Survey of Canada, Miscellaneous Reports* 43:1–31.

Darwin, C. 1859. *On the origin of species.* London: John Murray.

Darwin, C. 1868. *The variation of animals and plants under domestication.* 2 vols. London: John Murray.

Durham, J. W. 1974. Systematic position of *Eldonia ludwigi* Walcott. *Journal of Paleontology* 48:750–55.

Dzik, J., and K. Lendzion. 1988. The oldest arthropods of the East European platform. *Lethaia* 21:29–38.

Erwin, D. H., J. W. Valentine, and J. J. Sepkoski. 1987. A comparative study of diversification events: The early Paleozoic versus the Mesozoic. *Evolution* 141:1177–86.

Gagnier, P.-Y., A. R. M. Blieck, and G. Rodrigo. 1986. First Ordovician vertebrate from South America. *Geobios* 19:629–34.

Glaessner, M. F. 1984. *The dawn of animal life.* Cambridge: Cambridge University Press.

Gould, S. J. 1977. *Ever since Darwin.* New York: W. W. Norton.

Gould, S. J. 1981. *The mismeasure of man.* New York: W. W. Norton.

Gould, S. J. 1985a. The paradox of the first tier: An agenda for paleobiology. *Paleobiology* 11:2–12.

Gould, S. J. 1985b. Treasures in a taxonomic wastebasket. *Natural History Magazine* 94 (December):22–33.

Gould, S. J. 1986. Evolution and the triumph of homology, or why history matters. *American Scientist,* January–February, pp. 60–69.

Gould, S. J. 1987a. Life's little joke. *Natural History Magazine* 96 (April):16–25.

Gould, S. J. 1987b. Bushes all the way down. *Natural History Magazine* 96 (June):12–19.

Gould, S. J. 1987c. William Jennings Bryan's last campaign. *Natural History Magazine* 96 (November):16–26.

Gould, S. J. 1988. A web of tales. *Natural History Magazine* 97 (October):16–23.

Gould, S. J., N. L. Gilinsky, and R. Z. German. 1987. Asymmetry of lineages and the direction of evolutionary time. *Science* 236:1437–41.

Gould, S. J., D. M. Raup, J. J. Sepkoski, T. J. M. Schopf, and D. S. Simberloff. 1977. The shape of evolution: A comparison of real and random clades. *Paleobiology* 3:23–40.

Haeckel, E. 1866. *Generelle Morphologie der Organismen.* 2 vols. Berlin: Georg Reimer.

Hanson, E. D. 1977. *The origin and early evolution of animals.* Middletown, Conn.: Wesleyan University Press.

Hopson, J. A. 1977. Relative brain size and behavior in archosaurian reptiles. *Annual Review of Ecology and Systematics* 8:429–48.

Hou Xian-guang. 1987a. Two new arthropods from Lower Cambrian, Chengjiang, Eastern Yunnan [in Chinese]. *Acta Palaeontologica Sinica* 26:236–56.

Hou Xian-guang. 1987b. Three new large arthropods from Lower Cambrian, Chengjiang, Eastern Yunnan [in Chinese]. *Acta Palaeontologica Sinica* 26:272–85.

Hou Xian-guang. 1987c. Early Cambrian large bivalved arthropods from Chengjiang, Eastern Yunnan [in Chinese]. *Acta Palaeontologica Sinica* 26:286–98.

Hou Xian-guang and Sun Wei-guo. 1988. Discovery of Chengjiang fauna at Meishucun, Jinning, Yunnan [in Chinese]. *Acta Palaeontologica Sinica* 27:1–12.

Hughes, C. P. 1975. Redescription of *Burgessia bella* from the Middle Cambrian Burgess Shale, British Columbia. *Fossils and Strata* (Oslo) 4:415–35.

Hutchinson, G. E. 1931. Restudy of some Burgess Shale fossils. *Proceedings of the United States National Museum* 78(11):1–24.

Jaanusson, V. 1981. Functional thresholds in evolutionary progress. *Lethaia* 14:251–60.

Jablonski, D. 1986. Larval ecology and macroevolution in marine invertebrates. *Bulletin of Marine Science* 39:565–87.

Jefferies, R. P. S. 1986. *The ancestry of the vertebrates.* London: British Museum (Natural History).

Jerison, H. J. 1973. *The evolution of the brain and intelligence.* New York: Academic Press.

King, Stephen. 1987. *The tommyknockers.* New York: Putnam.

Kitchell, J. A., D. L. Clark, and A. M. Gombos, Jr. 1986. Biological selectivity of extinction: A link between background and mass extinction. *Palaios* 1:504–11.

Knoll, A. H., and E. S. Barghoorn. 1977. Archean microfossils showing cell division from the Swaziland System of South Africa. *Science* 198:396–98.

Lovejoy, A. O. 1936. *The great chain of being.* Cambridge, Mass.: Harvard University Press.

Ludvigsen, R. 1986. Trilobite biostratigraphic models and the paleoenvironment of the Burgess Shale (Middle Cambrian), Yoho National Park, British Columbia. *Canadian Paleontology and Biostratigraphy Seminars.*

Margulis, L. 1981. *Symbiosis in cell evolution.* San Francisco: W. H. Freeman.

Margulis, L., and K. V. Schwartz. 1982. *Five kingdoms.* San Francisco: W. H. Freeman.

Massa, W. R., Jr. 1984. *Guide to the Charles D. Walcott Collection, 1851–1940.* Guides to Collections, Archives and Special Collections of the Smithsonian Institution.

Matthew, W. D., and W. Granger. 1917. The skeleton of *Diatryma,* a gigantic bird from the Lower Eocene of Wyoming. *Bulletin of the American Museum of Natural History* 37:307–26.

Mikulic, D. G., D. E. G. Briggs, and J. Kluessendorf. 1985a. A Silurian soft-bodied fauna. *Science* 228:715–17.

Mikulic, D. G., D. E. G. Briggs, and J. Kluessendorf. 1985b. A new exceptionally preserved biota from the Lower Silurian of Wisconsin, USA. *Philosophical Transactions of the Royal Society, London* B 311:75–85.

Morris, R. 1984. *Time's arrows.* New York: Simon and Schuster.

Müller, K. J. 1983. Crustacea with preserved soft parts from the Upper Cambrian of Sweden. *Lethaia* 16:93–109.

Müller, K. J., and D. Walossek. 1984. Skaracaridae, a new order of Crustacea from the Upper Cambrian of Västergötland, Sweden. *Fossils and Strata* (Oslo) 17:1–65.

Murchison, R. I. 1854. *Siluria: The history of the oldest known rocks containing organic remains.* London: John Murray.

Parker, S. P. (ed.). 1982. *McGraw-Hill synopsis and classification of living organisms*. 2 vols. New York: McGraw-Hill.

Pflug, H. D. 1972. Systematik der jung-präkambrischen Petalonamae. *Paläontologische Zeitschrift* 46:56–67.

Pflug, H. D. 1974. Feinstruktur und Ontogenie der jungpräkambrischen Petalo-Organismen. *Paläontologische Zeitschrift* 48:77–109.

Raup, D. M., and S. J. Gould. 1974. Stochastic simulation and evolution of morphology—towards a nomothetic paleontology. *Systematic Zoology* 23(3):305–22.

Raup, D. M., S. J. Gould, T. J. M. Schopf, and D. S. Simberloff. 1973. Stochastic models of phylogeny and the evolution of diversity. *Journal of Geology* 81(5):525–42.

Rigby, J. K. 1986. Sponges of the Burgess Shale (Middle Cambrian) British Columbia. *Palaeontographica Canada*, no. 2.

Robison, R. A. 1985. Affinities of *Aysheaia* (Onychophora) with description of a new Cambrian species. *Journal of Paleontology* 59:226–35.

Romer, A. S. 1966. *Vertebrate paleontology*. 3d ed. Chicago: University of Chicago Press.

Rozanov, A. Yu. 1986. Problematica of the Early Cambrian. In A. Hoffman and M. H. Nitecki (eds.), *Problematic fossil taxa*, pp. 87–96. New York: Oxford University Press.

Runnegar, B. 1987. Rates and modes of evolution in the Mollusca. In K. S. W. Campbell and M. F. Day, *Rates of evolution*, pp. 39–60. London: Allen and Unwin.

Schidlowski, M. 1988. A 3,800-million-year isotopic record of life from carbon in sedimentary rocks. *Nature* 333:313–18.

Schopf, T. J. M. 1978. Fossilization potential of an intertidal fauna: Friday Harbor, Washington. *Paleobiology* 4:261–70.

Schuchert, C. 1928. Charles Doolittle Walcott (1850–1927). *Proceedings of the American Academy of Arts and Sciences* 62:276–85.

Seilacher, A. 1984. Late Precambrian Metazoa: Preservational or real extinctions? In H. D. Holland and A. F. Trendall (eds.), *Patterns of change in earth evolution*, pp. 159–68. Berlin: Springer-Verlag.

Sepkoski, J. J., R. K. Bambach, D. M. Raup, and J. W. Valentine. 1981. Phanerozoic marine diversity and the fossil record. *Nature* 293:435.

Simonetta, A. M. 1970. Studies of non-trilobite arthropods of the Burgess Shale (Middle Cambrian). *Palaeontographica Italica* 66 (n.s. 36):35–45.

Simpson, G. G. 1980. *Splendid isolation: The curious history of South American mammals*. New Haven: Yale University Press.

Størmer, L. 1959. Trilobitoidea. In R. C. Moore (ed.), *Treatise on invertebrate paleontology*, Part O. Arthropoda I, pp. 23–37.

Stürmer, W., and J. Bergström. 1976. The arthropods *Mimetaster* and *Vachonisia* from the Devonian Hunsrück Shale. *Paläontologische Zeitschrift* 50:78–111.

Stürmer, W. and J. Bergström. 1978. The arthropod *Cheloniellon* from the Devonian Hunsrück Shale. *Paläontologische Zeitschrift* 52:57–81.

Sun Wei-guo and Hou Xian-guang. 1987a. Early Cambrian medusae from Chengjiang, Yunnan, China [in Chinese]. *Acta Palaeontologica Sinica* 26:257–70.

Sun Wei-guo and Hou Xian-guang. 1987b. Early Cambrian worms from Chengjiang, Yunnan, China: *Maotianshania* Gen. Nov. [in Chinese]. *Acta Palaeontologica Sinica* 26: 299–305.

Taft, W. H., *et al.* 1928. Charles Doolittle Walcott: Memorial meeting, January 24, 1928. *Smithsonian Miscellaneous Collections* 80:1–37.

Valentine, James W. 1977. General patterns in Metazoan evolution. In A. Hallam (ed.),

Patterns of evolution. New York: Elsevier Science Publishers.

Vine, Barbara [Ruth Rendell]. 1987. *A fatal inversion.* New York: Bantam Books.

Vonnegut, Kurt. 1985. *Galápagos.* New York: Delacorte Press.

Walcott, C. D. 1891. The North American continent during Cambrian time. In *Twelfth Annual Report, U.S. Geological Survey,* pp. 523–68.

Walcott, C. D. 1908. Mount Stephen rocks and fossils. *Canadian Alpine Journal* 1(2):232–48.

Walcott, C. D. 1910. Abrupt appearance of the Cambrian fauna on the North American continent. Cambrian Geology and Paleontology, II. *Smithsonian Miscellaneous Collections* 57:1–16.

Walcott, C. D. 1911a. Middle Cambrian Merostomata. Cambrian Geology and Paleontology, II. *Smithsonian Miscellaneous Collections* 57:17–40.

Walcott, C. D. 1911b. Middle Cambrian holothurians and medusae. Cambrian Geology and Paleontology, II. *Smithsonian Miscellaneous Collections* 57:41–68.

Walcott, C. D. 1911c. Middle Cambrian annelids. Cambrian Geology and Paleontology, II. *Smithsonian Miscellaneous Collections* 57:109–44.

Walcott, C. D. 1912. Middle Cambrian Branchiopoda, Malacostraca, Trilobita and Merostomata. Cambrian Geology and Paleontology, II. *Smithsonian Miscellaneous Collections* 57:145–228.

Walcott, C. D. 1916. Evidence of primitive life. *Annual Report of the Smithsonian Institution for 1915* [published in 1916], pp. 235–55.

Walcott, C. D. 1918. Appendages of trilobites. Cambrian Geology and Paleontology, IV. *Smithsonian Miscellaneous Collections* 67:115–216.

Walcott, C. D. 1919. Middle Cambrian Algae. Cambrian Geology and Paleontology, IV. *Smithsonian Miscellaneous Collections* 67:217–60.

Walcott, C. D. 1920. Middle Cambrian Spongiae. Cambrian Geology and Paleontology, IV. *Smithsonian Miscellaneous Collections* 67:261–364.

Walcott, C. D. 1931. Addenda to description of Burgess Shale fossils, [with explanatory notes by Charles E. Resser]. *Smithsonian Miscellaneous Collections* 85:1–46.

Walcott, S. S. 1971. How I found my own fossil. *Smithsonian* 1(12):28–29.

Walter, M. R. 1983. Archean stromatolites: evidence of the earth's earliest benthos. In J. W. Schopf (ed.), *Earth's earliest biosphere: Its origin and evolution,* pp. 187–213. Princeton: Princeton University Press.

White, C. 1799. *An account of the regular gradation in man, and in different animals and vegetables.* London: C. Dilly.

Whittington, H. B. 1971. Redescription of *Marrella splendens* (Trilobitoidea) from the Burgess Shale, Middle Cambrian, British Columbia. *Geological Survey of Canada Bulletin* 209:1–24.

Whittington, H. B. 1972. What is a trilobitoid? In *Palaeontological Association Circular, Abstracts for Annual Meeting,* p. 8. Oxford.

Whittington, H. B. 1974. *Yohoia* Walcott and *Plenocaris* n. gen., arthropods from the Burgess Shale, Middle Cambrian, British Columbia. *Geological Survey of Canada Bulletin* 231:1–21.

Whittington, H. B. 1975a. The enigmatic animal *Opabinia regalis,* Middle Cambrian, Burgess Shale, British Columbia. *Philosophical Transactions of the Royal Society, London* B 271:1–43.

Whittington, H. B. 1975b. Trilobites with appendages from the Middle Cambrian, Burgess Shale, British Columbia. *Fossils and Strata* (Oslo) 4:97–136.

Whittington, H. B. 1977. The Middle Cambrian trilobite *Naraoia*, Burgess Shale, British Columbia. *Philosophical Transactions of the Royal Society, London* B 280:409–43.

Whittington, H. B. 1978. The lobopod animal *Aysheaia pedunculata* Walcott, Middle Cambrian, Burgess Shale, British Columbia. *Philosophical Transactions of the Royal Society, London* B 284:165–97.

Whittington, H. B. 1980. The significance of the fauna of the Burgess Shale, Middle Cambrian, British Columbia. *Proceedings of the Geologists' Association* 91:127–48.

Whittington, H. B. 1981a. Rare arthropods from the Burgess Shale, Middle Cambrian, British Columbia. *Philosophical Transactions of the Royal Society, London* B 292:329–57.

Whittington, H. B. 1981b. Cambrian animals: Their ancestors and descendants. *Proceedings of the Linnean Society* (New South Wales) 105:79–87.

Whittington, H. B. 1985a. *Tegopelte gigas*, a second soft-bodied trilobite from the Burgess Shale, Middle Cambrian, British Columbia. *Journal of Paleontology* 59:1251–74.

Whittington, H. B. 1985b. *The Burgess Shale.* New Haven: Yale University Press.

Whittington, H. B., and D. E. G. Briggs. 1985. The largest Cambrian animal, *Anomalocaris*, Burgess Shale, British Columbia. *Philosophical Transactions of the Royal Society, London* B 309:569–609.

Whittington, H. B., and S. Conway Morris. 1985. *Extraordinary fossil biotas: Their ecological and evolutionary significance.* London: Royal Society. Published originally in *Philosophical Transactions of the Royal Society, London* B 311:1–192.

Whittington, H. B., and W. R. Evitt II. 1953. *Silicified Middle Ordovician trilobites.* Geological Society of America Memoir 59.

Zhang Wen-tang and Hou Xian-guang. 1985. Preliminary notes on the occurrence of the unusual trilobite *Naraoia* in Asia [in Chinese]. *Acta Palaeontologica Sinica* 24:591–95.

Credits

1.1 Copyright 1940 by Charles R. Knight. Reproduced by permission of Rhoda Knight Kalt.

1.2 Copyright © Janice Lilien. Originally published in *Natural History* magazine, December 1985.

1.3 From Charles White, *An Account of the Regular Gradation in Man . . .* , 1799. Reprinted from *Natural History* magazine.

1.4 Reprinted by permission of Charles Scribner's Sons, an imprint of Macmillan Publishing Company, from Henry Fairfield Osborn, *Men of the Old Stone Age.* Copyright 1915 by Charles Scribner's Sons; copyright renewed 1943 by A. Perry Osborn.

1.7 Reprinted courtesy of the *Boston Globe.*

1.8 Reprinted courtesy of the *Boston Globe.*

1.9 Reprinted courtesy of Bill Day, *Detroit Free Press.*

1.11 Reprinted courtesy of Guinness Brewing Worldwide.

1.12 Reprinted courtesy of Granada Group PLC.

1.15 From James Valentine, "General Patterns in Metazoan Evolution," in *Patterns of Evolution,* ed. A. Hallam. Elsevier Science Publishers (New York). Copyright © 1977.

1.16(A) From David M. Raup and Steven M. Stanley, *Principles of Paleontology,* 2d ed. Copyright © 1971, 1978 W. H. Freeman and Company. Reprinted with permission.

1.16(B) Figure 4.6 in Harold Levin, *The Earth Through Time.* Copyright © 1978 by Saunders College Publishing, a division of Holt, Rinehart and Winston, Inc. Reprinted by permission of the publisher.

1.16(C) From J. Marvin Weller, *The Course of Evolution.* McGraw-Hill Book Co., Inc. Copyright © 1969.

1.16(E) From Robert R. Shrock and William H. Twenhofel, *Principles of Invertebrate Paleontology.* McGraw-Hill Book Co., Inc. Copyright © 1953.

1.16(F) From Steven M. Stanley, *Earth and Life Through Time*, 2d ed. Copyright © 1986, 1989 W. H. Freeman and Company. Reprinted with permission.

2.4, 2.5, 2.6 Smithsonian Institution Archives, Charles D. Walcott Papers, 1851–1940 and undated. Archive numbers SA-692, 89-6273, and 85-1592.

3.1 By permission of the Smithsonian Institution Press, from *Smithsonian Miscellaneous Collections*, vol. 57, no. 6. Smithsonian Institution, Washington, D.C.

3.3, 3.4, 3.5, 3.6, 3.7 From D. L. Bruton, 1981. The arthropod *Sidneyia inexpectans*, Middle Cambrian, Burgess Shale, British Columbia. *Philosophical Transactions of the Royal Society, London* B 295: 619–56.

3.8 From H. B. Whittington, 1978. The lobopod animal *Aysheaia pedunculata* Walcott, Middle Cambrian, Burgess Shale, British Columbia. *Philosophical Transactions of the Royal Society, London* B 284:165–97.

3.9, 3.10, 3.11 From D. L. Bruton, 1981. The arthropod *Sidneyia inexpectans*, Middle Cambrian, Burgess Shale, British Columbia. *Philosophical Transactions of the Royal Society, London* B 295: 619–56.

3.13, 3.14, 3.15, 3.16 From H. B. Whittington, 1971. Redescription of *Marrella splendens* (Trilobitoidea) from the Burgess Shale, Middle Cambrian, British Columbia. *Geological Survey of Canada Bulletin* 209:1–24.

3.17, 3.19 From H. B. Whittington, 1974. *Yohoia* Walcott and *Plenocaris* n. gen., arthropods from the Burgess Shale, Middle Cambrian, British Columbia. *Geological Survey of Canada Bulletin* 231:1–21.

3.20 From H. B. Whittington, 1975. The enigmatic animal *Opabinia regalis*, Middle Cambrian, Burgess Shale, British Columbia. *Philosophical Transactions of the Royal Society, London* B 271:1–43.

3.22 Reprinted by permission of Cambridge University Press.

3.23 From A. M. Simonetta, 1970. Studies of non-trilobite arthropods of the Burgess Shale (Middle Cambrian). *Palaeontographica Italica* 66 (n.s. 36):35–45.

3.24, 3.25, 3.26 From H. B. Whittington 1975. The enigmatic animal *Opabinia regalis*, Middle Cambrian, Burgess Shale, British Columbia. *Philosophical Transactions of the Royal Society, London* B 271:1–43.

3.27 From C. P. Hughes, 1975. Redescription of *Burgessia bella* from the Middle Cambrian Burgess Shale, British Columbia. *Fossils and Strata* (Oslo) 4:415–35. Reproduced with permission.

3.30 From S. Conway Morris, 1977. A new entoproct-like organism from the Burgess Shale of British Columbia. *Palaeontology* 20:833–45.

3.33 From S. Conway Morris, 1977. A redescription of the Middle Cambrian worm *Amiskwia sagittiformis* Walcott from the Burgess Shale of British Columbia. *Paläontologische Zeitschrift* 51:271–87.

3.35 From S. Conway Morris, 1977. A new metazoan from the Cambrian Burgess Shale, British Columbia. *Palaeontology* 20:623–40.

3.36 From D. E. G. Briggs, 1976. The arthropod *Branchiocaris* n. gen., Middle Cambrian, Burgess Shale, British Columbia. *Geological Survey of Canada Bulletin* 264:1–29.

3.37 From D. E. G. Briggs, 1978. The morphology, mode of life, and affinities of *Canadaspis perfecta* (Crustacea: Phyllocarida), Middle Cambrian, Burgess Shale, British Columbia. *Philosophical Transactions of the Royal Society, London* B 281:439–87.

3.39, 3.40(A–C) From H. B. Whittington, 1977. The Middle Cambrian trilobite *Naraoia*, Burgess Shale, British Columbia. *Philosophical Transactions of the Royal Society, London* B 280:409–43.

3.42, 3.43 From H. B. Whittington, 1978. The lobopod animal *Aysheaia pedunculata* Walcott, Middle Cambrian, Burgess Shale, British Columbia. *Philosophical Transactions of the Royal Society, London* B 284:165–97.

3.44 From D. E. G. Briggs, 1981. The arthropod *Odaraia alata* Walcott, Middle Cambrian, Burgess Shale, British Columbia. *Philosophical Transactions of the Royal Society, London* B 291:541–85.

3.47, 3.50 From H. B. Whittington, 1981. Rare arthropods from the Burgess Shale, Middle Cambrian, British Columbia. *Philosophical Transactions of the Royal Society, London* B 292:329–57.

3.51, 3.52, 3.53 From D. L. Bruton and H. B. Whittington. 1983. *Emeraldella* and *Leanchoilia*, two arthropods from the Burgess Shale, British Columbia. *Philosophical Transactions of the Royal Society, London* B 300:553–85.

3.55 From D. E. G. Briggs and D. Collins, 1988. A Middle Cambrian chelicerate from Mount Stephen, British Columbia. *Palaeontology* 31:779–98.

3.56, 3.57, 3.59 From S. Conway Morris, 1985. The Middle Cambrian metazoan *Wiwaxia corrugata* (Matthew) from the Burgess Shale and *Ogygopsis* Shale, British Columbia, Canada. *Philosophical Transactions of the Royal Society, London* B 307:507–82.

3.60, 3.61 From D. E. G. Briggs, 1979. *Anomalocaris*, the largest known Cambrian arthropod. *Palaeontology* 22:631–64.

3.63, 3.64 From H. B. Whittington and D. E. G. Briggs, 1985. The largest Cambrian animal, *Anomalocaris*, Burgess Shale, British Columbia. *Philosophical Transactions of the Royal Society, London* B 309:569–609.

3.65 From S. Conway Morris and H. B. Whittington, 1985. Fossils of the Burgess Shale. A national treasure in Yoho National Park, British Columbia. *Geological Survey of Canada, Miscellaneous Reports* 43:1–31.

3.67, 3.68, 3.69(A–B), 3.70 From H. B. Whittington and D. E. G. Briggs, 1985. The largest Cambrian animal, *Anomalocaris*, Burgess Shale, British Columbia. *Philosophical Transactions of the Royal Society, London* B 309:569–609.

3.73, 3.74 From D. E. G. Briggs and H. B. Whittington, 1985. Modes of life of arthropods from the Burgess Shale, British Columbia. *Transactions of the Royal Society of Edinburgh* 76:149–60.

4.1, 4.2, 4.3 Smithsonian Institution Archives, Charles D. Walcott Papers, 1851–1940 and undated. Archive numbers 82-3144, 82-3140, and 83-14157.

5.3 Drawing by Charles R. Knight: neg. no. 39443, courtesy of Department of Library Services, American Museum of Natural History.

5.5 Courtesy of A. Seilacher.

5.6 From R. C. Moore, C. G. Lalicker, and A. G. Fischer. *Invertebrate Fossils.* McGraw-Hill Book Co., Inc. Copyright 1952.

5.7 From A. Yu. Rozanov, "Problematica of the Early Cambrian," in *Problematic Fossil Taxa*, ed. Antoni Hoffman and Matthew H. Nitecki. Copyright © 1986 by Oxford University Press, Inc. Reprinted by permission.

Index